D1350542

Studies in Sociology

Edited by
PROFESSOR W. M. WILLIAMS
University College, Swansea

12
INTRODUCING SOCIAL STATISTICS

STUDIES IN SOCIOLOGY

1 THE SOCIOLOGY OF INDUSTRY
by S. R. Parker, R. K. Brown, J. Child and M. A. Smith

4 THE FAMILY
by C. C. Harris

5 COMMUNITY STUDIES
An Introduction to the Sociology of the Local Community
by Colin Bell and Howard Newby

6 SOCIAL STRATIFICATION
An Introduction
by James Littlejohn

7 THE STRUCTURE OF SOCIAL SCIENCE
A Philosophical Introduction
by Michael Lessnoff

9 THE SOCIOLOGY OF LEISURE
by Stanley Parker

10 A SOCIOLOGY OF FRIENDSHIP AND KINSHIP
by Graham A. Allan

11 THE SOCIOLOGY OF WOMEN
An Introduction
by Sara Delamont

INTRODUCING SOCIAL STATISTICS

Richard Startup
Lecturer in Sociology, University College, Swansea

Elwyn T. Whittaker
Lecturer in Statistics, University College, Swansea

London
GEORGE ALLEN & UNWIN
Boston Sydney

**George Allen & Unwin (Publishers) Ltd,
40 Museum Street, London WC1A 1LU, UK**

George Allen & Unwin (Publishers) Ltd,
Park Lane, Hemel Hempstead, Herts HP2 4TE, UK

Allen & Unwin Inc.,
9 Winchester Terrace, Winchester, Mass 01890, USA

George Allen & Unwin Australia Pty Ltd,
8 Napier Street, North Sydney, NSW 2060, Australia

First published in 1982

British Library Cataloguing in Publication Data

Startup, Richard
Introducing social statistics. – (Studies in sociology; 12)
1. Sociology – Statistical methods
I. Title II. Whittaker, Elwyn T.
III. Series 519.5'024309 HM24
ISBN 0-04-310012-0
ISBN 0-04-310013-9 pbk

Library of Congress Cataloging in Publication Data

Startup, Richard.
Introducing social statistics.
(Studies in sociology; 12)
Includes bibliographical references and index.
1. Social sciences–Statistical methods. I. Whittaker, Elwyn T. II. Title. III. Series: Studies in sociology (Allen & Unwin); 12.
HA29.S782 300'.72 81-10986
ISBN 0-04-310012-0 AACR2
ISBN 0-04-310013-9 (pbk.)

Set in 10 on 11 point Press Roman by Alden Press Ltd,
Oxford, London and Northampton
and printed in Great Britain
by Billing and Sons Ltd, Guildford,
London and Worcester

Contents

Preface *page* ix
1 Statistics and the Social Sciences 1
2 Frequency Distributions, Graphic Representation and Measures of Central Tendency 8
3 Measures of Variation and the Standard Normal Distribution 26
4 Probability 45
5 Sampling and Estimation 69
6 Hypothesis Testing with Single Samples 89
7 Statistical Inference with Two Samples 104
8 An Introduction to Analysis of Variance 121
9 The Chi-Square Test and Contingency Problems with Nominal Scales 131
10 Regression and Correlation 147
11 Sources of Statistics 166
References 184
Appendix: Statistical Tables 185
Answers to Exercises 196
Index 199

Preface

This book is intended primarily for social science students, who require an introduction to statistics which combines an account of basic ideas with an illustration of techniques. In recent years there has been a considerable increase in the use of statistics in all the major social subjects: sociology, social policy, social history, political science and education. Nevertheless, most students entering upon courses in further and higher education still lack a sufficient mathematical background to gain ready access to the rapidly advancing technical literature on quantitative methods. What is needed is a text which enables them to obtain a secure understanding of basic statistical procedures without the assumption of a high degree of mathematical sophistication. The intention here is to provide a book suitable for a first course in statistics which presupposes rather limited mathematical expertise (roughly corresponding to GCE O level), but conveys the fundamental ideas underlying both descriptive and inferential statistics.

In teaching statistical applications, there is frequently a problem in motivating students. The fear of mathematics is one difficulty, and another is the fact that statistics is sometimes approached simply and solely as a tool used in various fields which are themselves the primary source of interest. The danger is that statistics may be seen as a necessary evil. To ease the problem, one can stress the precise and logical nature of statistical ideas and argument which may appeal directly to students and is not infrequently contrasted by them, once their understanding increases, with less exact practices in other subjects. Again, at a different level, the choice of illustrations is of great importance. In this case examples have been selected which are of primary interest to sociologists, although data have also been derived from neighbouring disciplines, such as social psychology, political science, social work and education. Even when simplified examples are presented, they are based on realistic data. Statistics can be exhibited as an essential part of the equipment of a social scientist, but it is a refreshing and invigorating subject in its own right.

In a statistics book the sequencing of topics is important, particularly so as to facilitate the passage to the more difficult concepts of inferential statistics. Chapter 1 places statistics in a wider perspective, while also examining the basic issue of the types of data generated in the social field. Chapter 2 brings together the presentation of data, in tabular and pictorial forms, and measures of central tendency. In Chapter 3 a consideration of measures of variation leads on to the introduction of the standard normal distribution. Experience shows that beginning students can really come to grips with the standard deviation via its use in work on the normal curve. There follows a chapter on the basic ideas of probability.

This is a topic which is often found difficult, but it is certainly needed as a foundation for an understanding of statistical inference. In this case a systematic development with many explanations and illustrations is completed by the introduction of the binomial and Poisson distributions. The development of statistical inference occupies Chapters 5–9, the emphasis being on those methods (both parametric and non-parametric) which are of the greatest practical usefulness in the non-economic social studies. Chapter 10 describes bivariate regression and correlation. At all stages the details of necessary calculations are made explicit but indications are sometimes also provided as to how computation can be more readily achieved, using the cheap calculating machines which are now so frequently available. Interestingly, it is our experience that work with calculators can assist the understanding of arithmetical and statistical operations even though it may undermine ability at mental arithmetic. Finally, in Chapter 11, in the interests of completeness, an account is given of some of the more important – mainly official – sources of statistics.

It is a pleasure to express our appreciation to those colleagues who have offered suggestions and criticisms which have influenced the writing of this book. We should especially like to thank Professor A. G. Hawkes and Mr Richard Wadman, for their help and encouragement. We are indebted to Mrs Dianne Mowat, who typed the final manuscript so carefully and also to Mrs Pauline Dugmore and Mrs Sheila Goodall, for their great assistance. Thanks are also due to Dr H. R. Neave, of Nottingham University, who has allowed us to reprint (sometimes in simplified form) our Tables A, B, E, F and G in the Appendix from his *Statistics Tables for Mathematicians, Engineers, Economists and the Behavioural and Management Sciences*. The tables in Chapter 11 are reproduced from *Social Trends 8*.

R. S.
E. T. W.

Chapter 1

Statistics and the Social Sciences

The term 'statistics' is used in three senses. In everyday language it is usually taken to mean – in the words of the *Concise Oxford Dictionary* 'numerical facts systematically collected'. It may also refer – as in the title of this book – to the methods of collecting, classifying and analysing quantitative data (i.e. information). Or, it may be the plural of 'statistic' which is a technical term, the purpose of which is explained on p. 3, referring to a characteristic of a sample which has been selected. In this book the methods of dealing with data concerned with people and their relationships are at the centre of attention. However, before these are described we need to establish two things, first, the purpose of statistical methods, and secondly, the kinds of data generated in the social field.

The Functions of Statistics

Statistical methods are used for two main purposes: to describe or summarise data in the most suitable way, and to make inferences from such data. As far as the former is concerned, the social scientist is often confronted with large quantities of data and has to know how to proceed to make sense of them. This is the case, for instance, when a social survey has been carried out, and it is also true of information obtained from official sources such as the Census of Population. Essentially, we need to change the form of the data so that their meaning can more easily be grasped. This can be done by calculating measures such as percentages and averages which summarise the data. We might also represent the data in tabular or pictorial form in order to take advantage of the increased impact of these forms. Thus, if we were interested in incomes, we might calculate average income and a measure of how all the incomes were spread out about that average value. We might also produce a table showing the numbers of people with incomes between specified limits, as is done in most official publications. We could then produce a picture of this table, so that the distribution of incomes could be observed more easily.

Where our interest lies in the relationships between aspects such as income, occupation, housing and social mobility, then there is again a need to summarise data. For the simplest case of two items the correlation coefficient (p. 154) which specifies the strength of the association between them, is a useful summarising measure, and patterns of correlation lend themselves well to graphical presentation. However, descriptive statistics is even more necessary when the relationship between more than two aspects is being examined, since in undigested form the data would be very difficult to absorb.

The second function of statistics, which will in fact be the subject of the major part of this book, is to make estimates of, and draw inferences about, the characteristics of populations on the basis of information obtained from samples which have been drawn from those populations according to a specific criterion. In statistical language the term *population* does not necessarily refer to people, but is used to describe the complete set of things – individuals, social groups, relationships, objects or observations in which we are interested and for which we wish our results to apply, e.g. all the households in a city. A *sample*, on the other hand, is simply a part (or sub-set) of the population. Undoubtedly, the use of methods of statistical inference has been a great help in the development of the social sciences. There are several reasons for this, the most obvious of which is essentially practical.

The sociologist seeks to describe and analyse social relationships and group life in industrial societies. However, the modern nation state incorporates millions of inhabitants linked by a vast network of relationships. Considerations of time and money dictate that the social investigator cannot contact every resident of the country (or even of a city) in order to study a phenomenon of interest, e.g. marital breakdown. What is possible, though, is for him to specify precisely the category of persons or social groups, e.g. families and households, about which he wishes to generalise (the population), and then draw a – perhaps relatively small – sample from it. He can then proceed to investigate the phenomenon in appropriate depth by contacting the members of the sample. He may find, for instance, that within the sample, marital breakdown has occurred more often for those couples where the partners come from diverse social backgrounds. However, since the primary interest is in the defined population and not simply in the sample, there is a need to make certain inferences from the characteristics of the latter to those of the former. This can be achieved with the help of statistical theory. Perhaps the most widely known examples where this type of inference is involved are provided by opinion polls which seek to indicate the voting intentions of the nation's electorate on the basis of information from samples of approximately 2,000 adults.

Sometimes the situation is rather different in that the social scientist wishes to investigate the causal processes which may have generated the data. Yet the aim still will often be to generalise, so the procedures of inferential statistics are again required. In experimental research, for instance, the intention is to draw conclusions about relationships which apply under similar conditions. Again, in non-experimental studies the social scientist may examine some, or even all, of the available cases but nevertheless seek to generalise more widely. For instance, perusal of the criminal statistics for England and Wales reveals that crime rates tend to be higher in urban than in rural areas, but inferential statistics are needed if one is to conclude that this association may be expected to apply generally in similar circumstances. Of course, vagueness here surrounds the notion of 'similar circumstances', and it is a basic task of the sociologist and criminologist in advancing from the initial statistical evidence to clarify this idea and provide understanding of the link between the determinants of crime and the differing features of urban and rural life.

This type of example can help to indicate the differing ways in which statistical procedures are used in social subjects as compared with disciplines such as physics and chemistry. The fact is that in these latter fields experimentation is the basic method, but for a variety of reasons it is less often able to be used in the study of social phenomena. The sociologist most often seeks to analyse either data derived from naturally occurring situations (as is the case with official statistics or the findings of direct observation), or those generated by social surveys. In these cases there are typically many relevant factors and statistical procedures are needed to analyse their relative influence. For instance, crime rates are no doubt affected not simply by urban and rural conditions, but also by the effectiveness of local police forces, local sentencing policies and other factors (some unrelated to the urban-rural dimension). What may initially appear to be a simple relationship often proves to be highly complex and difficult to interpret. Though procedures of statistical inference are used in virtually all scientific fields (e.g. to handle problems of measurement error), the need for them in the social field is particularly acute.

Once the broad nature of statistical inference is clarified, it becomes easier to specify in a preliminary way the two types of problem with which it deals: *the estimation of population parameters using sample statistics*, and *tests of statistical hypotheses*. In connection with the first topic, the notion of a *parameter* refers simply to some measurable characteristic of a population (e.g. a proportion or average). On the other hand, a *statistic* is defined as a quantity calculated from the raw sample data (again a proportion or average would be an example) which may be used to estimate a parameter. In a study of overcrowding in a city the population might be defined as all dwellings within an identifiable geographical boundary. The parameter to be estimated might be the proportion of those dwellings which are overcrowded, according to a specific criterion. In proceeding to estimate this parameter, the social scientist might select a sample of dwellings in the city from a suitable list or map. He would then determine by direct investigation the proportion of overcrowded dwellings in the sample. The latter quantity would be the sample statistic. If the sample had been appropriately selected, with the help of statistical theory he would then be in a position to estimate the parameter. This usually involves specifying a range of values (perhaps centred on the statistic) and attaching a high degree of confidence to the claim that, if its value became known, the parameter would be found to lie within that range. Summarising, one can say that the investigator is able to estimate the required quantity, but the fact that only partial coverage of the population has been achieved leads to there being a certain calculable error – the sampling error.

The second type of problem indicated above – the testing of statistical hypotheses – can be illustrated as follows. Suppose in a study of political attitudes and voting behaviour, one wishes to determine whether the distribution of opinions is the same among the men and the women resident in a city. To this end a sample is chosen by appropriate means from the electoral roll and the political views of those selected are determined, e.g. the proportions intending to vote Conservative, Labour, and so on. Almost inevitably, one is likely to find some (possibly slight) aggregate differences in the patterns of responses of the men and women sampled. The basic issue, though, if differences are present, is whether these are simply a feature of the particular sample selected – which may be expected to be absent in alternative samples – or can be taken as a firm indication that there are real differences in the population itself. The decision as to which of these alternatives to accept is made following a 'statistical test of significance' (see Chapter 6). Among other things the decision depends upon the sample size and the magnitude of the differences within the sample.

Examples such as these may help to make clear just how useful descriptive and inferential statistical procedures can be in the analysis of empirical data in the social field. However, it is necessary to stress that statistical issues do not arise simply at the final stages of an investigation. It is very important that statistical considerations be brought in at the planning stage of such projects, especially in relation to the precise definition of the population to be studied; the question of whether to draw any samples and, if so, the method of drawing them and their size; and the procedures of estimation and/or testing which will be used. It is essential that decisions on these issues be taken before rather than after the start of data collection.

Types of Data

Though the logic of statistical argument is always the same, the frequency with which particular measures and techniques are used varies considerably depending upon the field of application. This is chiefly because there is a tendency for differing kinds of data to be generated in different fields, and it is an essential fact that the data which are collected substantially determine the statistical techniques which may be used. The significance of this point becomes clearer when the types of data generated in the social field are reviewed and in certain respects contrasted with those available elsewhere.

Nominal Scales

In all systematic study the most basic operation is classification. The aim is to sort the various elements – people, relationships, groups or whatever they may be – into categories that are as homogeneous as possible. Numbers or other symbols are used simply as names for the various categories. A *nominal scale* exists provided the categories are such that no element appears in more than one category and all elements are included. Examples are the classification of people by sex or religion (where there is usually a need for a category for those with no religion). This is also described as classification according to an *attribute* where this latter term refers essentially to a qualitative non-numerical description.

Ordinal Scales

Sometimes it proves possible not simply to define categories, but also to rank them according to the extent to which they possess a certain characteristic. This type of classification is, in fact, particularly common in the social sciences. In sociology data on socio-economic status are frequently presented in the form of a categorisation of persons (or families) into a number of ranked classes. The upper-, middle- and working-class distinction is particularly familiar. However, this is sometimes refined into a six-class model in which those in the upper-upper class are judged to be of higher prestige than those in the lower-upper class, who in their turn are higher than the upper-middle class, and so on. Additional instances of this type of scaling are provided by situations where people have been ranked according to such attributes as their popularity or the extent of their influence within a group. For an *ordinal scale* to exist, one needs both the conditions for a nominal scale, and also the property that where two elements are not classified together it is possible to say which of them possesses a specified characteristic (e.g. prestige or popularity) to a greater extent. However, with ordinal scaling, one cannot speak of the magnitudes of differences between elements, and mathematical operations such as addition and multiplication are not generally applicable.

Interval and Ratio Scales

Measurement in the everyday sense of this term is used to refer to situations where we are able not only to rank items in respect of a certain characteristic, but also to assign numerical values and determine the distances (or differences in magnitude) between them. This applies to physical data concerning weight, length and temperature and, in the social field, it is also true for money. An *interval scale* possesses the property that items can be measured in terms of a common and constant unit, e.g. weight in grams, temperature in degrees centigrade and money in pounds sterling. The further distinction is made in the literature between an interval scale and the somewhat higher level of measurement known as a *ratio scale*. The latter is defined as an interval scale with the additional feature that it possesses a 'true zero point' as its origin. The zero measurements for weight, length and money are 'true zero points' in this sense, recording the fact that there is no weight, length or money, but, contrastingly, the usual zero points for temperature measurement (i.e. 0°C or 0°F) are arbitrary. The ratio scale has the advantage that measurements can be compared directly, as when we say that one man has twice the income of another. This type of scale permits the use of all the ordinary mathematical operations and, therefore, is ideal for statistical purposes. However, almost every interval scale is anyway either a ratio scale, or it can be extended into one (as can temperature scales by the introduction of the absolute zero of temperature), so once a unit of measurement has been established, one can in practice feel confident that virtually all mathematical operations may be performed.

Unfortunately, measurement at an interval level is less often achieved in the social than in the physical sciences. In addition to the case of financial data already noted and the measurement of time (as in the recording of an age distribution), perhaps the commonest situations in the social sciences when interval scaling is achieved are those where there has been enumeration of some kind – of people or groups – and figures have been standardised by reference to a base population. Census publications are a ready source of this type of data, for one is told, for instance, the percentage of households in a locality which possess basic amenities, e.g. a fixed bath or shower.

The characteristics recorded in interval and ratio scales are perhaps most often referred to as variables, to be contrasted with the attributes classified in nominal scales. A *variable* is essentially an aspect which is subject to quantitative measurement and to which a numerical value is assigned, e.g. income or age. In connection with variables a further distinction relevant to statistical work is that between those which are discrete and those which are continuous. Roughly speaking, *discrete variables* take only certain values (often whole numbers), whereas *continuous variables* take on all possible values in some range. For example, household size is a discrete variable, since a household can only contain a whole number of persons. On the other hand, age is a continuous variable, since that quantity could be shown to take any value (less than approximately 113 years) provided the measurement accuracy were sufficiently precise (e.g. it could in principle be determined to the nearest second). The distinction between discrete and continuous data is relevant among other things to the way figures are displayed, for it would clearly be absurd to describe some households as containing between 4.5 and 5.5 persons as opposed to saying they consist of five persons, whereas in categorising persons by age, it is absolutely necessary to group together those with ages which fall in a specified range. Some variables (e.g. money) are essentially discrete, since there is a minimum unit (a halfpenny) which cannot be divided; yet that unit is so small that for most purposes data (e.g. on income) may be treated as continuous.

In connection with both descriptive and inferential statistics there is a need to assess the type of scale being used, and in the case of variables, to observe whether they are discrete or continuous. However, it is also worth noting that the three main scales possess a useful cumulative property. The ordinal scale has the features of a nominal scale together with an ordering relation. The interval scale is ordinal but it also possesses a unit of measurement. The consequence of this cumulative property is that one can always 'drop back' a level in analysing data. For instance, supposing one were uncertain whether some achievement scores obtained from schoolchildren constituted interval level scaling, then they could appropriately be dealt with as ranked data. Indeed, uncertainties of this type are by no means uncommon in the social field and the net result is that statistical procedures involving ranking become particularly important. However, the process of (as it were) 'downgrading' data should not be pursued in a cavalier fashion. This is because – if data genuinely do lie on an interval scale – one is ignoring, or in a sense 'throwing away' information, to treat them simply as ordinal. For instance, if exact income data are provided for a sample and these are then reduced to an income rank order, all information on the relative sizes of incomes is lost (e.g. that one individual's income is three times that of another).

In practice most of the pressure which is experienced by the social scientist is in the other direction, i.e. to assume a higher level of scaling than is warranted. As will become clearer in later chapters, the reason for this is that, if one can justify the assumption of interval or ratio scaling, more statistical techniques and (in a sense) more powerful ones become available. This places one in a better position to estimate quantities precisely and conclude that statistical relationships between attributes and/or variables do indeed hold. However, unjustified assumptions will lead to error and an inability to reproduce results; hence there is a need for caution. Only when an understanding of statistical technique develops alongside a familiarity with types of social data, can statistics be of the greatest service to the social scientist.

Glossary

Population
Sample
Parameter
Statistic
Nominal scale
Ordinal scale
Interval scale
Ratio scale
Attribute
Variable
Discrete variable
Continuous variable

Chapter 2

Frequency Distributions, Graphic Representation and Measures of Central Tendency

One of the major problems in dealing with data in the social sciences is that very often there are so many of them. So many, in fact, that in their raw form it is almost impossible just to look at the data and gain very much in the way of understanding. Thus it becomes necessary to change the form of the data; first, to make them more presentable so that their nature can be more readily seen and understood; secondly, to obtain, where appropriate, certain measures which in themselves will describe or represent the body of the data very easily in a numerical form.

Frequency Distributions

The following data are drawn from the 10 per cent sample of the 1971 Population Census and deal with the variable of household size for two different enumeration districts in Swansea, Wales. The observations are the numbers of persons in each household. It may well be that our interest lies in comparing household sizes in the two districts. As the data stand it

Table 2.1 *Raw Data on Household Size*

							District A									
3	2	2	1	4	2	3	2	5	1	4	2	2	3	2	4	
1	3	2	2	3	5	6	2	2	3	4	1	5	3	1	2	
2	3	4	2	5	2	3	1	2	4	5	2	3	1	2	4	
4	3	1	5	2	4	3	1	4	3	4	5	2	4	3	5	
1	3	4	5	4	2	3	5	4	3	1						
							District B									
2	3	4	3	3	4	3	2	2	3	4	5	2	3	6	2	
1	3	2	4	2	4	5	3	2	5	4	6	2	2	3	3	
4	2	4	1	2	3	2	2	3	4	4	2	5	2	4	6	
3	2	2	3	4	4	3	3	2	4	2	3	4	3	3	2	
2	3	2	4	4	2	2	6	2	3	3	4	4	5	2	3	
4	5	3	4	3	4	5	2	4	3	4	5	3	3	4	3	4

is very difficult, if not impossible, to make any comparisons between them. The first step in attempting to understand the nature of the data is to create, for each district, a *frequency table.* This is done by counting how many one-person households there are, then how many two-person households, and so on. The results are then produced in tabular form, as Table 2.2.

Table 2.2 *Frequency Distributions of Household Size*

	District A		District B
Size of household	*Number of households (frequency)*	*Size of household*	*Number of households (frequency)*
1	11	1	2
2	21	2	28
3	17	3	29
4	15	4	26
5	10	5	8
6	1	6	4
Total	75	Total	97

We can illustrate our discrete distributions by means of line charts (Figure 2.1) or by bar charts (Figure 2.2). In *line charts*, lines are used to represent frequencies by their lengths, whereas *bar charts* consist of bars (rectangles) which are of equal width and whose lengths are proportional to the frequencies they represent. In this example we have adopted the most common practice, which is to indicate variable values on the horizontal (x) axis and to record frequencies by reference to the vertical (y) axis.

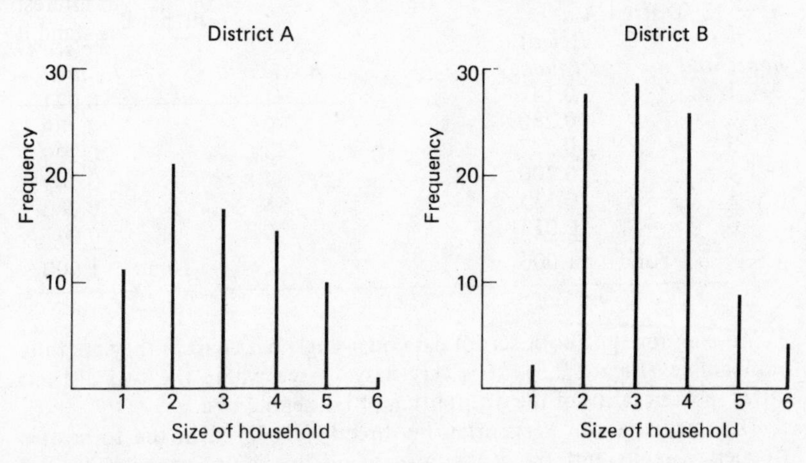

Figure 2.1 *Line charts.*

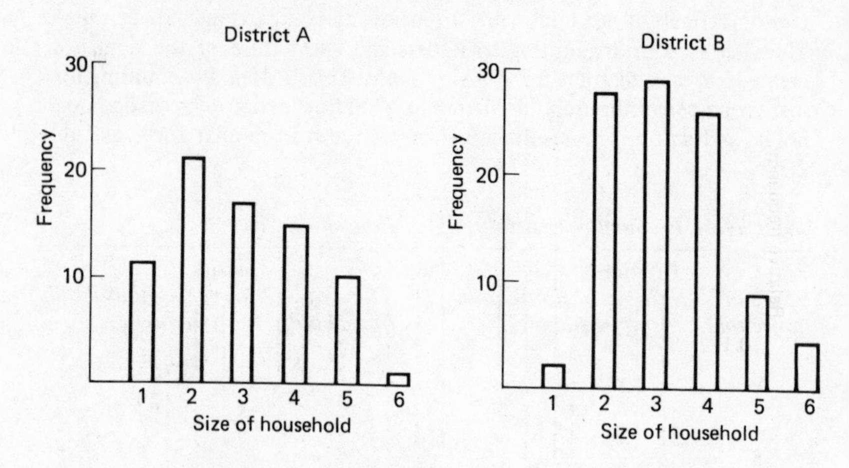

Figure 2.2 *Bar charts.*

Although quite a lot of information about the individual distributions can be obtained from Table 2.2 and Figures 2.1 and 2.2, we cannot compare them directly because the total frequency (number of households) is different for the two districts. To make them comparable, we must obtain *relative* (or proportionate) *frequencies*, i.e. the ratio frequency: total frequency, and this is given in Table 2.3.

Table 2.3 *Relative Frequency Distributions of Household Size*

District A		District B	
Size of household	*Relative frequency*	*Size of household*	*Relative frequency*
1	0.147	1	0.021
2	0.280	2	0.289
3	0.227	3	0.299
4	0.200	4	0.268
5	0.133	5	0.082
6	0.013	6	0.041
Total	1.000	Total	1.000

We can now put both sets of data on a single bar chart as they are fully comparable (Figure 2.3). It is very easy to see where the two districts differ most in terms of the distribution of household size.

Discrete data can very often be directly displayed in the form of a frequency table, but there are times when this is not appropriate. For example, if we consider the variable of school size as indicated by pupil

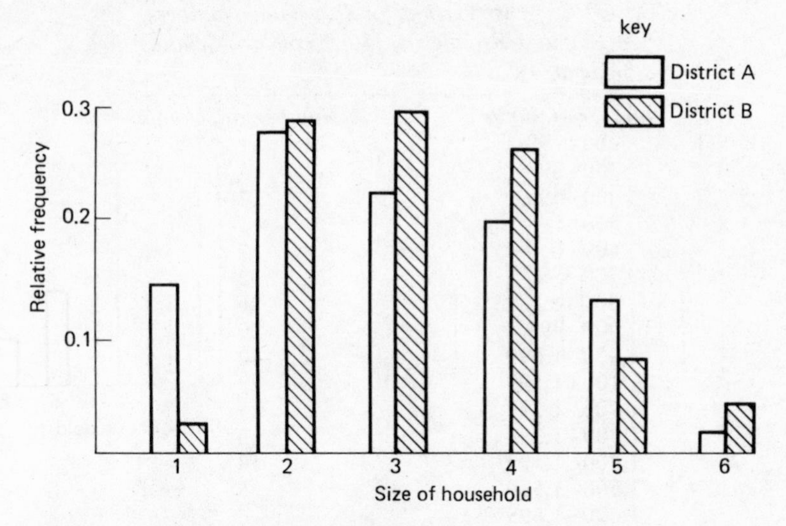

Figure 2.3 *Relative frequency bar chart.*

numbers, the individual values that this variable can take will be over a considerable range, so that a frequency table would not summarise them at all well. In such cases, sets of values are grouped together in the form of (say) schools with numbers of pupils from 200 to 299, from 300 to 399, and so on. We would then have a *grouped frequency table*, as in Table 2.4. Although such a table is informative, it possesses one major drawback compared with an ordinary frequency table. The sizes of the individual schools have been irretrievably lost. This feature will pose problems when, later in this chapter, we come to analyse tables of this kind.

To present our grouped frequency distribution graphically, use is made of a *histogram* (Figure 2.4). In a case like this where all the group intervals are of equal length, we proceed to construct the histogram by representing the measurements or observations constituting the set of data (in this case pupil numbers) on a horizontal scale and the group frequencies on the vertical scale. The graph is then formed by drawing rectangles, the bases of which are supplied by the group intervals and the heights of which are given by the corresponding group frequencies. It should be noted that for ease of presentation the rectangles of Figure 2.4 have been made to meet at the lower limits of the group intervals, i.e. 300, 400, etc., even though the group upper limits, i.e. 299, 399, etc., in fact differ from the adjacent lower limits (precisely because of the underlying discrete nature of the data). This procedure is satisfactory as long as we remain conscious of the true group limits, if we proceed to analyse the tabulated data further.

Continuous variable values are such that, except for small samples, they are almost always presented in the form of grouped frequency tables.

Table 2.4 *Distribution by Size (Pupil Numbers) of a Random Sample of 250 Secondary Schools in England, 1978*

Number of pupils	*Number of schools*
200–299	1
300–399	5
400–499	12
500–599	15
600–699	22
700–799	32
800–899	26
900–999	29
1,000–1,099	20
1,100–1,199	29
1,200–1,299	15
1,300–1,399	13
1,400–1,499	13
1,500–1,599	5
1,600–1,699	6
1,700–1,799	3
1,800–1,899	4
Total	250

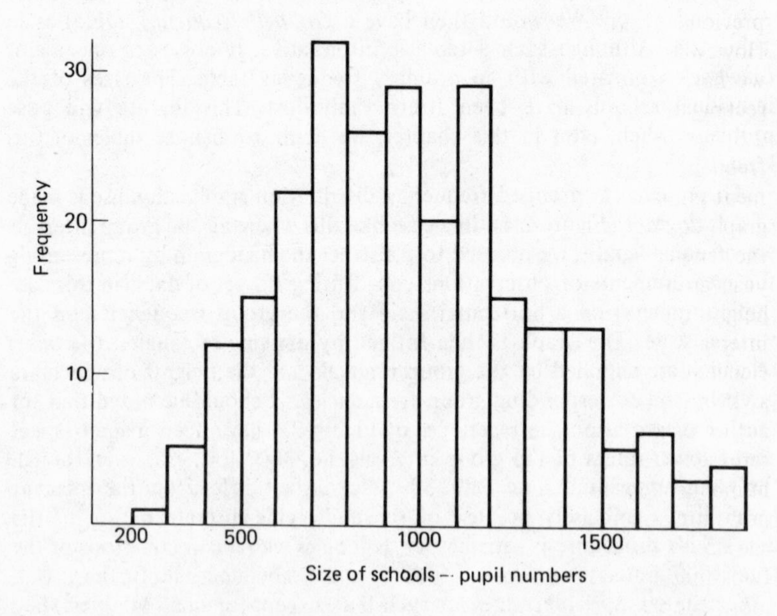

Figure 2.4 *Histogram – school sizes.*

The problem of having a large number of possible values, which we have just considered, is now compounded by there being no limitation at all on the values of the variable within the observed range. Quite often, values for some essentially continuous variables are given in a discrete form, and one must be careful in one's treatment of them. Age is a good example of a continuous variable frequently treated in this way. We talk about persons being 30 years of age, but they are 30 for an entire year until their next birthday when they become 31 years of age. Hence, it is evident that age is conventionally treated as being a discrete variable. From the 1971 Census returns for districts A and B, we have the grouped frequency distributions displayed in Table 2.5. One thing to notice here is the difference in the lengths of the groups. The first, 0–4, is five years in length; the next, 5–14, is ten; the next three groups are five; the next two are fifteen; the next five; and the last is what we call an open-ended group, i.e. there is no stated upper (or in other cases, lower) limit. One cannot say that this is an ideal way of grouping. In general, the less the group lengths, or intervals, differ, the better. In the above case there are good reasons why the Census returns should be grouped in such a way. Roughly they correspond to preschool children, those in compulsory education, students in further education or youngsters who have just left school, and so on. Once again, for purposes of comparison, it will help to convert the frequencies into relative frequencies (Table 2.6), and then draw histograms (Figure 2.5).

In constructing these histograms, we proceed in some respects as previously indicated for the grouped frequency distribution on school size. However, here we encounter two special problems of which the first is that we have unequal group intervals. If we ignored this fact and constructed rectangles with heights proportional to the relative frequencies of Table 2.6, we should tend to exaggerate the contribution made by the relative frequencies corresponding to the longer intervals. To avoid this, an adjustment is made. If the heights of the five-year groups are represented in the graph directly according to their relative frequencies, then for this purpose the ten-year group must have its relative frequency halved and the fifteen-year groups their relative frequencies divided by three. Essentially the height of the histogram corresponds to (relative frequency)/(length of interval), so that area under the histogram represents relative frequency. Hence, for each district the total area is one in Figure 2.5. The second awkward problem is how to portray the open-ended 65+ group. Some authors state that distributions with open-ended groups should never be represented in histograms, but this is perhaps too sweeping. Since the numbers of persons involved are not great, we have proceeded on the not unreasonable working assumption that the effective upper limit is age 85. From Table 2.6 and Figure 2.5, the main differences between the two districts are once again quite clear. Proportionately, there are far more people in the 0–4, 25–29 and 30–44 age groups in district B than district A, and far fewer in the 45–59, 60–64 and 65+ groups. These

Table 2.5 *Distributions by Age*

Age groups	*0–4*	*5–14*	*15–19*	*20–24*	*25–29*	*30–44*	*45–59*	*60–64*	*65+*	*Total*
District A	17	26	14	16	11	37	54	23	30	228
District B	51	41	14	20	45	90	37	5	10	313

Table 2.6 *Age Distributions: Relative Frequencies*

Age groups	*0–4*	*5–14*	*15–19*	*20–24*	*25–29*	*30–44*	*45–59*	*60–64*	*65+*	*Total*
District A	0.075	0.114	0.061	0.070	0.048	0.162	0.237	0.101	0.132	1.000
District B	0.163	0.131	0.045	0.064	0.144	0.288	0.118	0.016	0.032	1.001

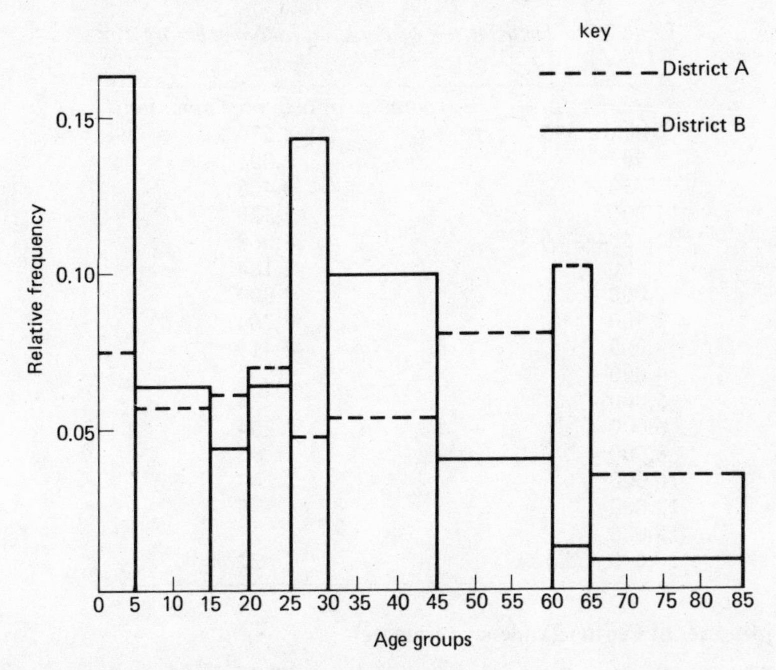

Figure 2.5 *Histogram – age distributions.*

observations provide important information concerning the types of population and areas with which we are dealing.

Observations on continuous variables must generally be put in the form of grouped frequency tables because it is most likely that very few observations will occur more than just a few times, many perhaps only occurring once. This same point applies to the essentially discrete but for most practical purposes continuous variable of income. Consider the data on annual income displayed in Table 2.7. Even with the very large frequencies of this table, if we exclude retired people with only their state pensions as income, probably not very many people (considered as a proportion of the total) would receive any individual indicated amount. The problem with a table like this is that one has to compromise between the amount of space to be taken up by the table and the amount of information to be imparted. If the group intervals were maintained at (say) £250, a great deal more information would be conveyed about the distribution, but the table would then occupy several pages. However, with the changing group intervals of the table in its existing form, one has to concentrate very carefully to obtain an accurate impression. We must stress that such tables are meant to be for presentation purposes only and are not readily amenable to further analysis. Wherever possible, calculations should be based on raw data values.

Table 2.7 *Distribution of Personal Incomes before Tax – UK 1973–4*

Income (£)	*Number of incomes (thousands)*
Below 595	3,579
595–	2,091
750–	3,408
1,000–	2,420
1,250–	2,068
1,500–	2,184
2,000–	3,800
2,500–	2,764
3,000–	2,415
4,000–	699
5,000–	234
6,000–	204
8,000–	85
10,000–	49
12,000–	38
15,000–	27
20,000 and over	22

Measures of Central Tendency (averages)

When, as is often the case, one is dealing with large numbers of observations on a particular attribute or variable, it is desirable, if not essential, to obtain measures which can be said to be representative of the data as a whole. Averages, or measures of central tendency, are one set which partially fulfils this purpose (others will be considered in Chapter 3).

The Mode

The simplest of the averages is the *mode*, which is the value occurring most frequently in a distribution. It is easy to obtain from a frequency table by observing which value has the highest frequency. For grouped data, it is more sensible to talk about the modal group (the group with the highest frequency) rather than the mode itself, for the latter can only be estimated at all precisely by the use of rather cumbersome formulae. The mode is the only appropriate average when data are recorded on a nominal scale. Table 2.8 contains illustrative data. The mode for both districts for the housing data is the owner-occupier group. For the car-ownership data the mode for district A is no car, whereas for district B it is one car.

The Median

The next simplest average is the *median*, which is defined as the middle value when the observations are arranged in order of magnitude, i.e. there are as many observations having values greater than the median as there are having values less than it. It is a suitable measure of central tendency when one is dealing with ordinally scaled data. For raw data, such as the

Table 2.8 *Data on Housing Tenure and Car Ownership*

	Number of Dwellings	
Housing Tenure	*District A*	*District B*
owner-occupier	44	94
renting from council	13	0
renting unfurnished	15	2
renting furnished	2	1
other/not stated	1	0
Total	75	97
	Number of Households	
Car Ownership	*District A*	*District B*
Households		
with no car	62	17
with 1 car	10	71
with 2 cars	3	9
Total	75	97

following, representing whooping cough deaths in England and Wales in the years 1960–72, the first step is to rearrange the observations in order of magnitude:

Year	(1960)	1961	1962	1963	1964	1965	1966	1967
No. of deaths	(38)	27	24	37	47	24	22	28
Year	1968	1969	1970	1971	(1972)			
No. of deaths	17	6	15	28	(2)			

If we initially focus on the 1961–71 period, these appear as

6 15 17 22 24 [24] 27 28 28 37 47

The middle value is 24, as there are five values above it and five below it. Here we have an odd number of observations, so that the median is one of those observations. What happens when there are an even number of observations? If we add the 1972 value to the data, the rearranged observations become

2 6 15 17 22 [24] [24] 27 28 28 37 47

There is no middle observation now, but as luck will have it, the two middle observations, i.e. the sixth and seventh, have the same value, so once again the median is 24. However, had we included the 1960 value in the series instead of the 1972 value, we would have had

6 15 17 22 24 [24] [27] 28 28 37 38 47

The two middle values are now 24 and 27, and there is no single value that we can call the median. All we can confidently assert is that the median lies on the interval from 24 to 27. In such cases it is conventional to take the value of the median as being halfway between the two middle values, i.e. 25.5 in this instance. When dealing with continuous variables, the convention is eminently reasonable for the resultant value is perfectly meaningful. However, with discrete variables conventional practice is rather less meaningful – as can be judged from our chosen example – for the value obtained for the median (25.5) may be such that it does not, and indeed cannot, occur in the data.

In determining the median from frequency distributions, it is helpful if one imagines the observations stretched out in sequence. For the frequencies shown in Table 2.2 and referring to district A, imagine eleven 1s followed by twenty-one 2s, seventeen 3s, fifteen 4s, ten 5s and one 6. There are seventy-five observations, so the median will be the thirty-eighth, i.e. the sixth of the seventeen 3s, all of which are indistinguishable from each other. The median will always be one of the values in a frequency distribution when the total N is odd and will almost always be when N is even (the exceptional case occurring when there are two different middle values).

Many of the variables we deal with are such that the highest frequencies occur near the middle of the distribution. Lengths of stay for in-patients in hospitals, Intelligence Quotient scores (IQs) and the sizes of schools as indicated by pupil numbers are examples of such variables. The median and the mode are called measures of central tendency (or location) because, for such distributions, their values are always near the centre of the distribution. As descriptive measures the median and the mode are quite useful and easy to understand, but they are somewhat limited in their application when we are dealing with interval-scaled variables. This brings us to the *arithmetic mean*, the last of the measures of central tendency which we shall discuss. This is what most people know as the 'average'.

The Arithmetic Mean

We have all heard of the term 'average' via the media and respond at least intuitively to the idea of the average wage, average height, average family size, and so on. In making this idea more precise, it is merely necessary to explain how the average is calculated. Nevertheless, many people still experience a need to know what the arithmetic mean 'is' as opposed to how it is calculated. To try to understand the nature of the arithmetic mean, imagine a weightless scale along which one hangs a unit weight for each observation of a particular value. If we use the whooping cough data (1961–71), then Figure 2.6 gives a representation of it in this form. Now consider a pointer or knife edge moving along the scale until at some value the system of weights is in balance. This value is the arithmetic mean. It is a point of balance for a set of values.

The calculations involved in finding the value of the (arithmetic) mean

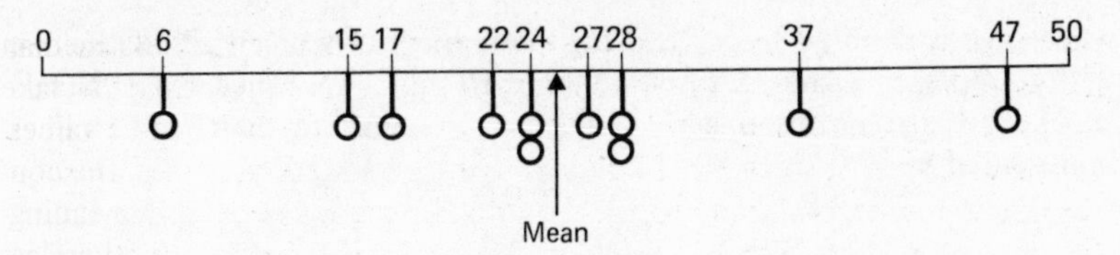

Figure 2.6 *The mean as a balancing point.*

are quite straightforward. One simply adds up all the values and divides by the number of observations. For the 1961–71 whooping cough data the total number of deaths over the eleven years was 275, so that the mean or average number of deaths per annum was $275/11 = 25$ for that period. However, if we now include the 1972 figure of two deaths, the value of the mean falls to $277/12 = 23.1$, and this serves to illustrate two of the problems associated with the mean. First, if the data are discrete the mean is not necessarily one of the observed values, nor even a value which the variable could possibly take. Secondly, it is very susceptible to extreme values, e.g. in our example the mean fell in value from 25 to 23.1 following the inclusion of just one extra value. We note, however, that the median was the same for both sets of data. This brings us to the question of which measure of central tendency we should use. As we have indicated above, with nominal and ordinal scales the choice is limited as the mean is only appropriate with intervally scaled variables. Ideally, however, where we have the latter, all three measures should be given for descriptive purposes, since together they can impart quite a lot of information.

The mean, unsupported by any other measure, is of only limited value, although one might not get this impression from the press and television. For example, the statement that the average income for people in the UK is now £4,000 per annum tells us nothing about the way that incomes are distributed, which for many purposes is far more important than the actual mean value. The important factor which must be taken into account, in order to get an accurate picture, is the variability of the data. If our average income of £4,000 is derived from a distribution in which everyone gets between £3,500 and £4,500, i.e. where there is little variation, then we have an egalitarian society. But if it is derived from a distribution with many incomes of less than £2,000 and a small number of perhaps £100,000 or more, then a very different type of society exists. We shall discuss variation, which is basic to all statistical method, fully in Chapter 3.

The Σ (sigma) Notation. At this point we must introduce a general method of representing both data and results. Let us suppose that we are interested in a particular variable, for example, the age at which women bear their first child. Since it is much more convenient in statistics to use symbols rather than words, we denote the age of the first woman at childbirth by x_1, the age of the second woman by x_2, and so on, up to the nth woman, whose age is denoted by x_n, n being the number of women whose ages we wish to represent. These n (as yet unknown) values are called the x-values.

The average or mean age at which n women had their first child is represented by

$$\bar{x} = \frac{x_1 + x_2 + x_3 + \ldots + x_n}{n} \tag{2.1}$$

that is, the sum of all the observed values divided by the number of values. On our way to determining x-values and hence calculating $\bar{x}$, we would generally proceed by defining the population under study (say, all the women with children in a defined area) and we might then select a sample of size n from it. The x-values would then be ascertained from the sample. The symbol $\bar{x}$ (x bar) is used to represent the mean of a sample of size n. The mean of a population of size N is represented by the Greek letter μ (mu).

The expression 2.1 for $\bar{x}$ is not very convenient if we have to write it repeatedly, so what is needed is a form of shorthand to replace it, hence the Σ notation representing summation. The expression

$$\bar{x} = \frac{\sum_{i=1}^{n} x_i}{n} \tag{2.2}$$

replaces that in expression 2.1 and, in words, says 'add up all the x-values starting with x_i with $i = 1$, i.e. x_1, and increasing i by one each time until you reach x_i with $i = n$, i.e. x_n; then divide the total by n'. The main advantages in using the Σ notation will become apparent in subsequent chapters.

The Arithmetic Mean for Frequency Distributions. Expression 2.2 is the one used in finding the mean for raw data. For example, using the raw data on household sizes in Table 2.1, we find that the mean for district A is 2.93, while for district B the mean is 3.23. We obtain these figures by determining that for district A the total number of people in all seventy-five households is 220, while for district B the total is 313 in ninety-seven households. Now let us look at the problem of finding the mean for the corresponding frequency distributions (see Tables 2.2 and 2.9). For district A we see that there are 11 households with only one person. These contribute 11 towards the total number of people. There are 21 households with two people, contributing 42 persons to the total; 17 households of three people contribute 51; 15 households of four contribute 60; 10 households of five contribute 50 and 1 household of six contributes 6. Adding up all the contributions, we have $11 + 42 + 51 + 60 + 50 + 6 = 220$, which as we would expect is exactly the same total number of people as we obtained from the raw data. We can also confirm that the total number of households – 75 – is the sum of the various numbers of households of each particular size. If we carry out this scheme in tabular form for both districts, we obtain the results shown in Table 2.9.

Table 2.9 *Calculation of Arithmetic Mean for Frequency Distributions*

Household size x_i	District A *Number of households* f_i	$f_i x_i$
1	11	11
2	21	42
3	17	51
4	15	60
5	10	50
6	1	6
	$\sum_{i=1}^{k} f_i = 75$	$\sum_{i=1}^{k} f_i x_i = 220$

Household size x_i	District B *Number of households* f_i	$f_i x_i$
1	2	2
2	28	56
3	29	87
4	26	104
5	8	40
6	4	24
	$\sum_{i=1}^{k} f_i = 97$	$\sum_{i=1}^{k} f_i x_i = 313$

If the variable value is represented by x_i, the associated frequency is f_i and there are k categories, then the sum of products of frequency and value is represented by $\sum_{i=1}^{k} f_i x_i$ in the Σ notation, while the sum of the frequencies is $\sum_{i=1}^{k} f_i$. Hence, for frequency distributions the mean is given by the expression:

$$\bar{x} = \frac{\sum_{i=1}^{k} f_i x_i}{\sum_{i=1}^{k} f_i} \tag{2.3}$$

Directly from the table one can calculate the two means as 2.93 and 3.23, respectively. For comparison the mode for district A is 2; for district B it is 3. The median for district A (the thirty-eighth observation) is 3; for district B (the forty-ninth observation) it is also 3.

The Arithmetic Mean for Grouped Frequency Distributions. Obtaining the mean for grouped frequency data, is not as straightforward as for frequency data. The main problem is that once an observation has been placed in a group it loses its individual identity. It is no longer possible (as was the

Table 2.10 *Calculation of Arithmetic Mean for Grouped Frequency Distributions*

District A

Age group	*Mid-point* x_i	*Frequency* f_i	$f_i x_i$
0–4	2.5	17	42.5
5–14	10.0	26	260.0
15–19	17.5	14	245.0
20–24	22.5	16	360.0
25–29	27.5	11	302.5
30–44	37.5	37	1,387.5
45–59	52.5	54	2,835.0
60–64	62.5	23	1,437.5
65 and over	75.0	30	2,250.0
		$\sum_{i=1}^{k} f_i = 228$	$\sum_{i=1}^{k} f_i x_i = 9{,}120.0$

District B

Age group	*Mid-point* x_i	*Frequency* f_i	$f_i x_i$
0–4	2.5	51	127.5
5–14	10.0	41	410.0
15–19	17.5	14	245.0
20–24	22.5	20	450.0
25–29	27.5	45	1,237.5
30–44	37.5	90	3,375.0
45–59	52.5	37	1,942.5
60–64	62.5	5	312.5
65 and over	75.0	10	750.0
		$\sum_{i=1}^{k} f_i = 313$	$\sum_{i=1}^{k} f_i x_i = 8{,}850.0$

If the group mid-point value is represented by x_i, the associated frequency by f_i and there are k categories, then the mean is again given by formula 2.3. So for district A,

$$\bar{x} = \frac{9{,}120.0}{228} = 40.0, \text{ and for district B, } \bar{x} = \frac{8{,}850.0}{313} = 28.3$$

Note: Many calculators have the facility to analyse frequency data as directly as they do raw data (obtaining $\sum_{i=1}^{k} f_i x_i$ and $\sum_{i=1}^{k} f_i$, simultaneously), so there would then be no need for the full tabular presentation of Tables 2.9 and 2.10.

case with frequency tables) to retrieve the original value. Let us look once again at the data on age distributions for districts A and B, given in Table 2.5. We no longer know the exact ages (even to the nearest year) of the seventeen children in district A who were under 5. All we know is that they had not reached their fifth birthday. While grouping is useful for presentation, it poses problems for analysis.

In order to find the mean, we shall have to make certain assumptions about the way in which the observations are spread over the group intervals. The simplest assumption, and the one we shall use, is that they are spread uniformly between the group limits. This amounts to treating the members of (say) the 20–24 group as if they had a mean age of 22.5 years. One must bear in mind that although the lower limit of the group, i.e. 20, is a well-defined fixed point – twenty years to the day, hour or minute after the person was born – the upper limit of the group may not appear to be so well defined. Essentially the group extends as near to 25 as we choose to go, if we treat age as the truly continuous variable that it is. So in practice, we assume that the upper limit actually is 25 and that the mid-point of the group is 22.5.

In this respect age is treated differently from many other continuous variables. For example, if we say that someone is 160 cm tall, we assume that this is a 'rounded' value, i.e. that the person's actual height lies between 159.5 cm and 160.5 cm. This meaning is not given to an age of (say) 30 years. In that case the person is more than 30 but less than 31 years of age, and for a number of people all 30 years old (in the absence of additional information), it is safest to proceed on the assumption that the mean age is 30.5. So, in general, for grouped data we treat all observations as having the mid-group value. Then we can carry out our calculations exactly as we did for frequency data, multiplying the mid-group values by the corresponding frequencies and summing the results (see Table 2.10).

A further problem arises in this instance as to how to treat the open-ended 65+ group. The upper limit is not specified and again we are forced to make an assumption. Where all the groups are of the same size, it may not be too unreasonable to assume that the open-ended group is the same size as the others. However, in our example, this does not apply. In fact, whatever assumption we make may well be erroneous. For district A about 14 per cent of people fall into this group and, therefore, our assumption will have a marked effect on the value of the mean. For district B only about 3 per cent are over 65, so the effect here will be much less. Probably the best procedure is to take the mid-point for the group as being 75, for district A follows a not dissimilar pattern to the national age distribution in which the mean age of the over-65s is about 75. The formal method of obtaining the mean for our two distributions is shown in Table 2.10.

A Comparison

We can now appropriately draw together all the information presented in this chapter regarding the two Swansea districts A and B and see if our findings can be integrated into a broader picture. From Table 2.3, we see that the districts differ most as far as household size is concerned in the frequency of one-person units (where A has relatively more) and in three- and four-person units (of which B has more). From Table 2.6, we note that the age distributions differ most in the 0–4, 25–29 and 30–44 groups (in which B has relatively more persons) and for those aged 45 and over (of which A has more). The housing and car-ownership data in Table 2.8 show very clear differences between the two districts. District B has almost 100 per cent home-owners and over 80 per cent of households with cars, whereas district A has only about 60 per cent home-owners and

80 per cent of households with no car. It should not now be too difficult for the reader to make his own deductions as to the types of population/area we are describing here.

The data on household size are not by themselves conclusive. The higher percentage of one-person households in district A may seem to suggest either lodgers or old people living on their own. The higher percentage in the three- and four-person households in district B may indicate a suburban type of area containing couples with one or two children. However, both of these suggestions are speculative. Useful supporting information comes from the age distributions, for district B has a much younger population than district A – only 17 per cent over 45 as compared with 47 per cent and twice as many under-5s. This is certainly suggestive of district B being a housing estate and district A being an older area nearer the city centre. The housing and car ownership data back up this view quite strongly. The very high proportion of owner-occupiers in district B together with the extensive car ownership are very indicative of a private suburban estate. The much more mixed tenure and the very low extent of car ownership in district A point to it being a relatively old and economically poor area near the urban centre. This analysis and discussion may serve to dramatise the point that statistics is not simply concerned with interpreting individual tables, but also with assembling evidence bearing on a particular topic or theme. We shall calculate some additional measures which summarise data on these two districts in Chapter 3.

Glossary

Frequency distribution
Line chart
Bar chart
Histogram
Relative (or proportionate) frequency
Grouped frequency distribution
Group interval
Mode
Median
Arithmetic mean
Σ notation

Exercises

1 Draw histograms for the following distributions:

(a)

Number of children in family x	Number of families f
0	26
1	42
2	68
3	61
4	29
5	10
6	3
8	1
Total	240

(b)

Marks obtained in sociology examination x	Number of students f
Under 40	3
40–44	4
45–49	15
50–54	67
55–59	63
60–64	48
65–69	3
70 and over	2
Total	205

2 The following are the population densities (persons/km^2) for all the major countries in the world. Lay out the data in the form of a grouped frequency table and draw a histogram:

6; 5; 127; 12; 3; 3; 3; 24; 20; 38; 16; 13; 19; 11; 1; 11; 37; 4; 1; 35; 10; 3; 60; 136; 20; 39; 4; 16; 13; 6; 14; 33; 31; 41; 20; 6; 26; 41; 41; 191; 79; 390; 3,829; 164; 81; 18; 22; 141; 280; 24; 268; 115; 323; 13; 70; 1; 67; 121; 123; 4; 3,527; 70; 45; 33; 133; 105; 29; 75; 88; 317; 77; 113; 114; 14; 93; 240; 150; 67; 111; 42; 178; 319; 12; 104; 105; 85; 67; 18; 152; 11; 228; 80; 9; 4; 11; 13; 19; 34; 73; 89; 21; 165; 48; 175; 23; 171; 25; 15; 19; 6; 11; 184; 16; 11; 2; 22; 306; 2; 7; 10

Calculate the mean population density from the raw data and from your grouped frequency distribution. Is the mean the best measure of central tendency to use here? If so, why? If not, indicate what is and calculate its value. Is the mean population density the same as the population density for these countries as a whole?

3 For the following Census data on household size (x) for three districts, A, B, and C, obtain in each case the mean, median and mode:

District A		District B		District C	
x	f	x	f	x	f
1	9	1	9	1	12
2	24	2	29	2	12
3	16	3	11	3	10
4	20	4	10	4	8
5	20	5	6	5	5
6	2	6	6	6	1
8	2	7	3	Total	48
Total	93	Total	74		

4 Look at the following sets of raw data and in each case tick those of the possible answers given that you think might be the correct value for the mean. Cross out those which cannot be correct. Then determine the correct value in each case:

	Mean?			
3; 5; 6; 10; 12; 14; 15; 17; 18; 20	5	15	12	21
12; 13; 13; 13; 14; 15; 15; 16; 16; 18	17	11	15	14.5
8.1; 8.5; 9.2; 9.4; 9.9; 10.3; 10.4; 10.7; 11.5; 12.0	8.6	10.0	9.5	11.7
1,030; 1,051; 1,083; 1,105; 1,127; 1,157; 1,189; 1,202; 1,256; 1,290	1,045	1,149	1,161	1,132

Chapter 3

Measures of Variation and the Standard Normal Distribution

The term *variable* which was used in the first two chapters implies that what we are interested in, whether it be household size, IQ, or income, is subject to variation. Households have varying numbers of members, people have different IQs, varying incomes and they are of two different sexes. These are examples of naturally occurring variation, and it is because of such variation that the subject of statistics is an important one.

The Range

In order to understand the nature of variation (or dispersion), let us first take a simple example of a variable on an interval scale. Two students, studying the same six subjects, had the following sets of examination marks:

	Economics	Geography	Sociology	Politics	Psychology	Statistics
Student A	51	55	48	57	58	49
Student B	69	47	46	48	42	66

If we add up the marks for each student, we find that the total is 318. So their mean or average mark is the same and equals 53. But do they have similar patterns of ability? We note that student A's lowest mark is 48 and his highest mark is 58, whereas student B's lowest mark is 42 and his highest mark is 69. The difference between highest and lowest values is termed the *range*, and it is the simplest measure of variation that we have:

$$\begin{aligned}\text{Student A: range} &= 58-48 = 10\\ \text{Student B: range} &= 69-42 = 27\end{aligned}$$

The range of marks for student B is much greater than for student A, and we might say that A was more consistent in his marks than B. Whereas B is very good at economics and statistics but rather poor at the other four subjects, A shows moderate ability in most if not all of his six subjects, although he is outstanding in none of them.

Although on some occasions the range may be useful in differentiating between grossly different sets of values, it is too elementary a measure to be of much general use. The main problem is that just one extreme observation can make an enormous difference to the value of the range even when we are dealing with very large numbers of observations. There is, therefore, a need for a more sophisticated measure.

The Quartile Deviation

A measure which goes some way to overcome the deficiency of the range is the quartile deviation. In Chapter 2 we saw that the median was the value which had equal numbers of observations above and below it when the data were arranged in order of magnitude. If instead we split our data into four equal parts, then the value below which one-quarter of the observations lie is called the first or lower quartile, designated Q_1; the value below which three-quarters of the observations lie is called the third or upper quartile, designated Q_3. In this context the median is the second quartile, Q_2. The *quartile deviation* Q is defined as being half the difference between the upper and lower quartiles, i.e.

$$Q = \frac{Q_3 - Q_1}{2} \tag{3.1}$$

This measure is not affected by extreme values, since the quartiles are well away from the extremities of the distribution. Quartile deviation is quite often used as a measure of variation when the median is being used as the measure of central tendency.

Deciles and Percentiles

Deciles may be thought of as an extension of the idea of quartiles. If we split the distribution (and we are really referring here to essentially continuous data) into ten equal parts, then the boundaries between the parts are called the *deciles*. The first decile is the value below which one-tenth of the observations lie, the second decile is the value below which two-tenths lie, and so on. The *percentiles* extend the idea further, in that the distribution is then split into 100 equal parts. The tenth and twentieth percentiles coincide with the first and second deciles. The 10 per cent of people with the lowest scores in (say) an IQ distribution are those with an IQ of less than the tenth percentile value. Decile and percentile values can materially aid the description of the variation in some observations.

The Mean Deviation

In the attempt to develop satisfactory measures of variation to be used with intervally scaled data, it is both convenient and desirable that these should use the arithmetic mean as a base value. In fact, the more important measures of variation make use of differences, or deviations, from the mean value. Let us look at these deviations, using the two sets of students' marks presented above:

Student A:	marks	51	55	48	57	58	49
	deviations	51−53 = −2	55−53 = +2	48−53 = −5	57−53 = +4	58−53 = +5	49−53 = −4
Student B:	marks	69	47	46	48	42	66
	deviations	69−53 = +16	47−53 = −6	46−53 = −7	48−53 = −5	42−53 = −11	66−53 = +13

In both cases it is seen that the sum of the deviations from the mean is zero. This is always the case, for in general we have

$$\begin{aligned}\text{sum of deviations from mean} &= \sum_{i=1}^{n} (x_i - \bar{x}) \\ &= (x_1 - \bar{x}) + (x_2 - \bar{x}) + (x_3 - \bar{x}) + \ldots + (x_n - \bar{x}) \\ &= (x_1 + x_2 + x_3 + \ldots + x_n) - n\bar{x} \\ &= \sum_{i=1}^{n} x_i - n \sum_{i=1}^{n} x_i/n \\ &= 0\end{aligned}$$

As we noted in Chapter 1, the mean is a balancing value. The above result expresses this idea more formally, the mean having a value such that the sum of the deviations about it is zero.

This result poses a problem. Any measure of variation obviously cannot be based on the sum of the deviations in their raw form. They have to be modified somehow, so that the zero-sum is avoided. One way would be to ignore the information as to whether the observations are above or below the mean and concentrate on how many units they are away from the mean. All that this involves, is treating all deviations as being positive. In mathematical terms, we say we take the *modulus* of each of the deviations. If we then add up all these positive values and find their average, or mean, we obtain the measure known as the *mean deviation* (or *average deviation*). In the Σ notation we have

$$\text{mean deviation} = \frac{\sum_{i=1}^{n} |x_i - \bar{x}|}{n} \qquad (3.2)$$

where the upright lines are the symbols for treating the quantity within as being positive. These lines are not to be confused with parentheses (brackets).

To calculate the mean deviation for the marks of each student, all we need to do is to add the numerical values of the deviations from the mean and divide by the number of marks, i.e. 6:

for student A, mean deviation = 22/6 = 3.67
for student B, mean deviation = 58/6 = 9.67

One can note that in this case the ratio between the two mean deviations (2.64) is very close to the ratio between the ranges (2.70). This is certainly not so in all cases, although in many empirical examples they are not too different.

For frequency distributions, the only change in the calculations necessary to find the mean deviation is to multiply each modulus of the deviation from the mean by the frequency appropriate to the x-value with which one is dealing, then sum these products for all k categories and divide by the total frequency. More formally,

$$\text{mean deviation} = \frac{\sum_{i=1}^{k} f_i|x_i - \bar{x}|}{\sum_{i=1}^{k} f_i} \tag{3.3}$$

In our experience, it is not very often that one needs to make use of the mean deviation. Though it is a useful descriptive measure of variation, it suffers from two major defects. First, deviations taken irrespective of sign are not easily manipulated algebraically. Secondly, the mean deviation is not easily interpreted theoretically, so it is not very suitable for use in further statistical work. Part of its value here has been as a stepping-stone to help us reach the measures of variation which are theoretically superior.

The Variance and the Standard Deviation

If we are unable to use the moduli of the deviations from the mean, how else can the problem of the zero-sum be overcome? The mathematician will quickly provide the answer. Square the deviations, he will say, and then find the mean of these squared deviations. The quantity we thus obtain is called the *variance*, and it is the basic measure of variation in all but the most elementary statistical work. In the Σ notation, for raw data

$$\text{variance} = \frac{\sum_{i=1}^{n} (x_i - \bar{x})^2}{n} \tag{3.4}$$

In verbal terms it is the mean squared deviation (i.e the mean of the squares of the deviations). The variance is perhaps the most important measure in statistics, but there is one major problem that may be encountered in trying to use it descriptively. The original observations (the x-values) will have been in certain units – possibly years or pounds sterling or marks expressed as percentages. The arithmetic mean will also have been in these same units, and so will the deviations from the mean. But the variance, based on squared deviations, will be in units squared and thus will not be comparable with the original observations. Even though it has certain vital properties which will necessitate its use later in this book, for the time being we need a measure which is in the same units as our data, and to this end we take the square root of the variance. The new measure, called the *standard deviation*, and given the symbol S for sample data, is defined as

$$S = \sqrt{\frac{\sum_{i=1}^{n} (x_i - \bar{x})^2}{n}} \tag{3.5}$$

or, in words, the root mean squared deviation. For a population the standard deviation is referred to by the symbol σ (i.e. the Greek 's', sigma).

It is to be regretted that the variance has not been given a symbol unconnected with the standard deviation since, as things stand, the variance appears as σ^2 for populations and S^2 for samples and seems to be merely the square of the standard deviation, whereas it is in fact the more fundamental measure. However, at this stage, the standard deviation plays the more prominent role in our work, first, because it is in the same units as the observations and, secondly, because of its use in work on the normal distribution, which we shall deal with shortly.

The definition 3.5 of the standard deviation is not the most useful form as far as calculations are concerned. It would involve, first, adding up the observations and dividing by n to obtain the mean, calculating the deviations from the mean, squaring them, and then obtaining their sum. This would give the numerator expression and one would then divide by n and determine the square root. However, there is a much more useful computational form which can be very easily derived from the definition. First, take the numerator expression and expand it:

$$\sum_{i=1}^{n} (x_i - \bar{x})^2 = (x_1 - \bar{x})^2 + (x_2 - \bar{x})^2 + (x_3 - \bar{x})^2 + \ldots + (x_n - \bar{x})^2$$

$$= (x_1^2 - 2x_1\bar{x} + \bar{x}^2) + (x_2^2 - 2x_2\bar{x} + \bar{x}^2) + \ldots + (x_n^2 - 2x_n\bar{x} + \bar{x}^2)$$

We can now see that the first term in each bracket is the square of one of the x-values, the second term is always minus twice the x-value times the mean and the third term is the square of the mean. We next gather together all the first terms, then all the second terms and then all the third terms to obtain

$$(x_1^2 + x_2^2 + x_3^2 + \ldots + x_n^2) - (2x_1\bar{x} + 2x_2\bar{x} + 2x_3\bar{x} + \ldots + 2x_n\bar{x})$$
$$+ (\bar{x}^2 + \bar{x}^2 + \ldots + \bar{x}^2)$$

In the Σ notation the first bracket becomes $\sum_{i=1}^{n} x_i^2$. The second bracket becomes $2\bar{x}(x_1 + x_2 + x_3 + \ldots + x_n) = 2\bar{x}\sum_{i=1}^{n} x_i$. In the third bracket we are adding up $\bar{x}^2$ n times, so the result is $n\bar{x}^2$. Making the substitution $\bar{x} = \sum_{i=1}^{n} x_i/n$, we now have:

$$\sum_{i=1}^{n}(x_i-\bar{x})^2 = \sum_{i=1}^{n} x_i^2 - 2\left(\sum_{i=1}^{n} x_i\right)^2\Big/ n + \left(\sum_{i=1}^{n} x_i\right)^2\Big/ n$$

$$= \sum_{i=1}^{n} x_i^2 - \left(\sum_{i=1}^{n} x_i\right)^2\Big/ n$$

The value of this new expression for the sum of squares of deviations is that one need only accumulate the x-values and their squares and then find the difference between the two expressions shown above. The alternative computational formula for the standard deviation is

$$S = \sqrt{\frac{\sum_{i=1}^{n} x_i^2 - \left(\sum_{i=1}^{n} x_i\right)^2\Big/ n}{n}} \tag{3.6}$$

In this connection, one must be sure to note that the terms $\sum_{i=1}^{n} x_i^2$ and $(\sum_{i=1}^{n} x_i)^2$ are totally different. The first is the sum obtained by adding all the squares of the original values, while the second is the total of all the observations which is itself then squared. A further formula for calculating the standard deviation for raw data is

$$S = \sqrt{\frac{\sum_{i=1}^{n} x_i^2}{n} - \left(\frac{\sum_{i=1}^{n} x_i}{n}\right)^2} \tag{3.7}$$

which is obtained by separating the terms in expression 3.6. In this case the first term under the square root sign is the sum of the squares of the observations divided by the total number of observations, while the second term is the square of the mean, i.e. $\bar{x}^2$.

The computational formulae 3.6 and 3.7 are often useful whether or not one has access to a calculator. One problem with formula 3.5 is that it requires the preliminary determination of the mean, and it is difficult to judge at what degree of accuracy to determine this latter quantity if serious error in S is not to result from 'rounding' up or down the value of $\bar{x}$. Formulae 3.6 and 3.7 avoid this problem. Modern calculators vary in their capacities, because they are designed for differing uses. The most basic carry out only the elementary arithmetical operations. However, there are those with statistical facilities which can obtain directly the mean and standard deviation for raw data (and possibly frequency data as well). These calculators have a minimum of three memories which are needed to hold n, $\sum_{i=1}^{n} x_i$ and $\sum_{i=1}^{n} x_i^2$. The mean and standard deviation are obtained simply by pressing the appropriate buttons after the data have been entered. This sets the machine to calculate $\bar{x}$ and S, using expressions 2.2 and 3.6. These machines are ideal for our purposes.

If we now take the students' marks given at the beginning of the chapter, namely, student A, 51, 55, 48, 57, 58, 49, and student B, 69, 47, 46, 48, 42, 66, then from formula 3.6 the standard deviations are

$$\text{for student A} \quad S = \sqrt{\frac{16{,}944 - (318)^2_6}{6}} = 3.87$$

$$\text{and for student B} \quad S = \sqrt{\frac{17{,}510 - (318)^2_6}{6}} = 10.46$$

Note that these values are not very different from the previously determined values for the mean deviation. Generally, the mean deviation and the standard deviation are fairly close, with the former being the smaller quantity.

Having calculated the standard deviation for each set of marks, one may be inclined to ask what use can be made of it. Clearly, one can say that B's marks are much more variable than A's, but we might have deduced that from the marks themselves. What we do have is two values of a standardised quantitative measure which takes us beyond our subjective impressions. Nevertheless, it is apparent that the standard deviation will prove to be more useful descriptively in cases where there are large numbers of observations. Beyond this, one must assert that both the standard deviation and the variance are theoretical measures and their importance can only be revealed gradually. Beginning students are inclined to ask 'What is the standard deviation?', but besides saying it is a measure of variation, one can only add that its significance is revealed by its many uses.

Before proceeding to consider calculations on frequency tables, it becomes convenient to simplify the Σ notation somewhat. The reader should now be familiar with the idea that generally our summations have been taken over the values from x_1 to x_n, where n is a definite but unspecified whole number. Henceforward, where the context is clear, we shall use Σx in place of $\sum_{i=1}^{n} x_i$, and Σx^2 instead of $\sum_{i=1}^{n} x_i^2$. Other summations will generally be presented in this simpler form.

The Standard Deviation for Frequency Distributions

In frequency tables, we saw in Chapter 2 (p. 21) that if the variable value is represented by x_i, the associated frequency is f_i and there are k categories, then the mean $\bar{x}$ is given by $\sum_{i=1}^{k} f_i x_i / \sum_{i=1}^{k} f_i$, or, in our simplified notation, $\Sigma fx/\Sigma f$. As compared with the formula for the mean of raw data, we can say that Σfx replaces Σx, and Σf, the total frequency, replaces n. Similarly, in developing a formula for the standard deviation for a frequency distribution based on formula 3.7 above for raw data, we substitute Σfx for Σx, and Σf for n. It simply remains to consider what replaces Σx^2 in formula 3.7. Now where we have a number of observations with identical values, summing their squares involves adding the square of the value as many times as it appears, which is the same as multiplying the square of the value by the number of times it appears, i.e. by the frequency. If we have

Table 3.1 *Calculation of Standard Deviation for Frequency Distributions*

District A				District B			
Household size x	*Frequency* f	fx	fx^2	*Household size* x	*Frequency* f	fx	fx^2
1	11	11	11	1	2	2	2
2	21	42	84	2	28	56	112
3	17	51	153	3	29	87	261
4	15	60	240	4	26	104	416
5	10	50	250	5	8	40	200
6	1	6	36	6	4	24	144
	$\Sigma f = 75$	$\Sigma fx = 220$	$\Sigma fx^2 = 774$		$\Sigma f = 97$	$\Sigma fx = 313$	$\Sigma fx^2 = 1{,}135$

For district A, $\bar{x} = \Sigma fx/\Sigma f = 220/75 = 2.93$

variance $= S^2 = \Sigma fx^2/\Sigma f - (\Sigma fx/\Sigma f)^2 = 774/75 - (220/75)^2$

$= 10.320 - 8.604 = 1.716$

standard deviation $= \sqrt{\text{variance}} = S = 1.31$

For district B, $\bar{x} = \Sigma fx/\Sigma f = 313/97 = 3.23$

variance $= S^2 = \Sigma fx^2/\Sigma f - (\Sigma fx/\Sigma f)^2 = 1{,}135/97 - (313/97)^2$

$= 11.701 - 10.412 = 1.289$

standard deviation $= \sqrt{\text{variance}} = S = 1.13(5)$

Note: The easiest way of obtaining fx^2 values is by multiplying each fx value by the corresponding x-value.

(say) the value 4 appearing six times, the sum of squares will be $16 + 16 + 16 + 16 + 16 + 16 = 96 = 16 \times 6$. So the Σx^2 term will be replaced by Σfx^2, which is the sum of products of frequency and value squared for all values in the table. Thus, for frequency data,

$$S = \sqrt{\frac{\Sigma fx^2}{\Sigma f} - \left(\frac{\Sigma fx}{\Sigma f}\right)^2} \qquad (3.8)$$

The analysis of frequency tables is best done by retaining the tabulated form, although some calculators have the facility of accepting frequency data direct. Let us analyse in Table 3.1 the data on household size from Table 2.9 in Chapter 2. We can observe that district B has, on average, households with more occupants than district A, but has a smaller standard deviation, i.e. less variation in household size. We might have expected this because, for district B, 83 out of the 97 households (86 per cent) have sizes of two, three or four, i.e. within approximately one of the mean, whereas for district A, only 53 out of 75 (71 per cent) have sizes within approximately one of the mean. As our measures of variation are based on differences (deviations) from the mean, inevitably district B must show a smaller variation than district A. It is important that this kind of approach be adopted in calculations. One must always ask whether the results seem reasonable. Sometimes before a calculation, it is possible to obtain advance estimates of any required measures and this can also give greater confidence in results. For instance, we might expect the mean for district B to be just over three, because that particular value is the mode and the distribution is approximately symmetrical but with a slight excess of larger households. This kind of procedure should at least avoid gross errors, which can occur even with the most careful use of machines.

The Standard Deviation for Grouped Frequency Distributions

We stated in Chapter 2 that wherever practicable one should avoid calculating measures based on grouped data, and gave our reasons there. However, there will be occasions when such calculations cannot be avoided (perhaps because data only becomes available in that form), so we shall deal with the method for determining the standard deviation here. Essentially, our method is an extension of that described in Chapter 2 for calculating the mean. This involves taking the mid-points of the groups as if they were values in a frequency table, and then using formula 3.8. As an illustration, in Table 3.2 we analyse the data on the age distributions for our two Swansea districts provided in Table 2.10.

As we noted in Chapter 2 the average age for district B is considerably less than that for district A, and we are now in a position to state that the standard deviation is also smaller, although both standard deviations are quite large. Determining that our results are reasonable or getting advance estimates of our measures, is quite difficult here because of the changing group intervals. If we standardise the group intervals so that with the

Table 3.2 *Calculation of Standard Deviation for Grouped Frequency Distributions*

District A

Age group	*Mid-point* x	*Frequency* f	fx	fx^2
0–4	2.5	17	42.5	106.25
5–14	10.0	26	260.0	2,600.00
15–19	17.5	14	245.0	4,287.50
20–24	22.5	16	360.0	8,100.00
25–29	27.5	11	302.5	8,318.75
30–44	37.5	37	1,387.5	52,031.25
45–59	52.5	54	2,835.0	148,837.50
60–64	62.5	23	1,437.5	89,843.75
65 and over	75.0	30	2,250.0	168,750.00
		$\Sigma f = 228$	$\Sigma fx = 9{,}120.0$	$\Sigma fx^2 = 482{,}875.00$

$$\bar{x} = \frac{\Sigma fx}{\Sigma f} = \frac{9{,}120}{228} = 40.0$$

$$\text{variance } S^2 = \frac{\Sigma fx^2}{\Sigma f} - \left(\frac{\Sigma fx}{\Sigma f}\right)^2 = \frac{482{,}875}{228} - \left(\frac{9{,}120}{228}\right)^2 = 517.9$$

standard deviation $S = 22.8$

District B

Age group	*Mid-point* x	*Frequency* f	fx	fx^2
0–4	2.5	51	127.5	318.75
5–14	10	41	410.0	4,100.00
15–19	17.5	14	245.0	4,287.50
20–24	22.5	20	450.0	10,125.00
25–29	27.5	45	1,237.5	34,031.25
30–44	37.5	90	3,375.0	126,562.50
45–59	52.5	37	1,942.5	101,981.25
60–64	62.5	5	312.5	19,531.25
65 and over	75.0	10	750.0	56,250.00
		$\Sigma f = 313$	$\Sigma fx = 8{,}850.0$	$\Sigma fx^2 = 357{,}187.50$

$$\bar{x} = \frac{\Sigma fx}{\Sigma f} = \frac{8{,}850}{313} = 28.275$$

$$\text{variance } S^2 = \frac{\Sigma fx^2}{\Sigma f} - \left(\frac{\Sigma fx}{\Sigma f}\right)^2 = \frac{357{,}187.5}{313} - \left(\frac{8{,}850}{313}\right)^2 = 341.7$$

standard deviation $S = 18.5$

exception of the open-ended group they are of equal length, our tables appear thus:

District A		District B	
Age group	*Frequency*	*Age group*	*Frequency*
0–14	43	0–14	92
15–29	41	15–29	79
30–44	37	30–44	90
45–59	54	45–59	37
60 and over	53	60 and over	15

The mean for district A seems likely to be in the upper part of the 30–44 age group because of the greater pull of the over-45s as compared with the under-30s. The frequencies are not far from being uniform, with perhaps a slight tendency to being U-shaped (i.e. when presented in a histogram). This must make the measure of variation higher, since the largest deviations from the mean will be the most frequent. On the other hand, the mean for district B seems likely to be at the upper end of the 15–29 group or the lower end of the 30–44 group because of the very great pull of the under-30s. Although the proportion of people over 45 is not very large, their ages are some way removed from the mean so that, when combined with the large 0–14 group, the variation is relatively high despite the sizable 30–44 group. A method of obtaining advance estimates of the standard deviation will be dealt with below during the work on the normal distribution.

Accuracy and Approximation. This is an appropriate point to consider when one should or should not 'round' figures up or down and the kind of accuracy one should aim for. As far as the level of accuracy is concerned, when using a calculator there is a tendency to write down as the answer a number read off to six or even eight decimal places. This is almost always unnecessary and sometimes definitely wrong. Let us consider again our grouped frequency distributions and an appropriate level of accuracy for the mean and standard deviation. The first thing of which to remind ourselves is that information is lost through the grouping of Table 3.2, so that we cannot indicate individual ages even to the nearest year, let alone more accurately. This being so, it makes no sense to specify for descriptive purposes the mean or standard deviation to (say) two or more decimal places. Clearly, the accuracy of results is dictated by that of the information initially provided. However, in this case, we could reason that there is probably a tendency for errors due to grouping to cancel out when we sum a large number of observations or squares of observations, so it may be a reasonable compromise to state results – as we have done – to one decimal place. Certainly such figures are meaningful, since we are dealing here with a continuous variable. The information of Table 3.1 provides a contrast, because in that case the variable is discrete and no information has been lost through grouping. As a consequence in a purely technical sense the results are accurate to any number of decimal places, but are they meaningful when so presented? Since household size is necessarily a whole number, it is far more realistic when using the mean as a purely descriptive measure to say of the data in Table 3.1 that the mean household size for district A was just below 3 and for district B rather more than 3. The reason for usually recording the values of $\bar{x}$ and S to a greater degree of accuracy than this is basically because further analysis may then be conducted of the type which will be indicated in subsequent chapters.

The Normal Distribution

Having considered some basic summarising measures, we introduce a particularly important frequency distribution, the *normal (*or *Gaussian) distribution*. Not only do many empirical distributions approximate to the normal, but it is also of vital importance in inferential statistics, the topic of Chapters 5–9. So far the frequency distributions which we have discussed have involved a finite number of observations. However, it turns out to be useful in connection with statistical theory sometimes to think in terms of indefinitely large (or infinite) populations and to represent these graphically by means of curves. The normal distribution is one such, and it is represented by the *normal curve* which can be expressed by a mathematical equation. For our purposes here there is no need to consider the equation, but it will prove helpful to see how the distribution can be represented by a curve.

So far we are familiar with histograms of a finite aggregate of observations based on a limited number of group intervals. Suppose the number of equal intervals is initially three. Then imagine we double the number to six (by halving the length of each interval), while correspondingly increasing the number of cases. The (rough) pattern of the histogram is retained but it is more stepped (Figure 3.1(a)). We continue the process. At the next stage there are twelve group intervals, again with an increased number of cases (Figure 3.1(b)). Continuing further to twenty-four intervals and beyond, we note that the general pattern of the histogram is retained, while the number of rectangles increases. What we can say is that as the number of intervals increases, the histogram approximates more and more closely to a smooth curve. It is this curve which represents the (indefinitely large) population. For the normal distribution the curve (superimposed in Figure 3.1(b)) is symmetrical and bell-shaped.

The normal curve represents the associated distribution exactly. It approaches closely but does not touch the x-axis at any point (except at the mathematical notion of infinity). For this and other reasons it can only be approximated by empirical frequency distributions such as those of people's height or IQ. Given the distribution's symmetrical bell shape, its mean, median and mode all coincide. It has other interesting and remarkable properties. For instance, once its mean and variance (or standard deviation) are known, this enables one to determine precisely the proportion of observations lying between any two values (i.e. the proportion of the area under the curve between any two x-values). This property can be most useful when we know that the normal curve closely approximates another distribution which is of interest. When the population distribution of a variable x is exactly normal with mean μ and variance σ^2, we represent this by

$$x \sim N(\mu, \sigma^2) \qquad (3.9)$$

The symbol $\sim$ denotes 'distributed as', N tells us that the distribution is normal, while the first value inside the brackets indicates the mean of the

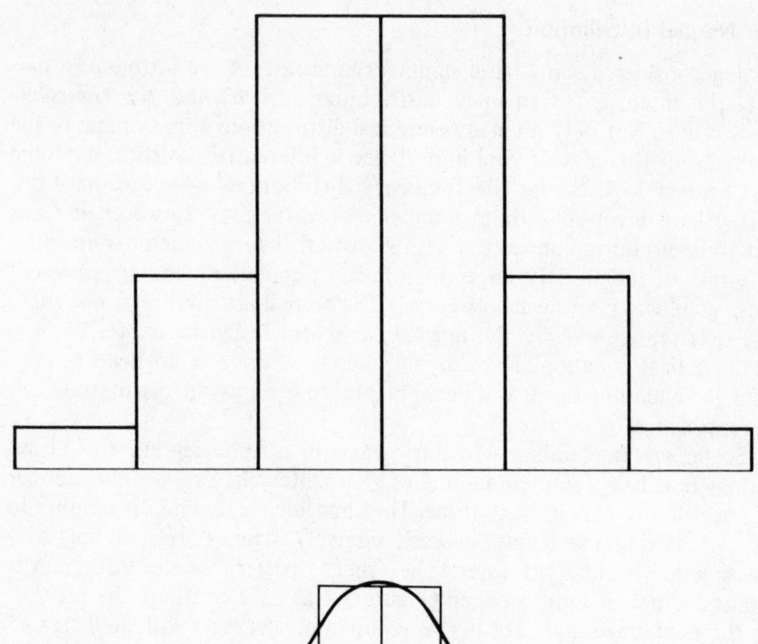

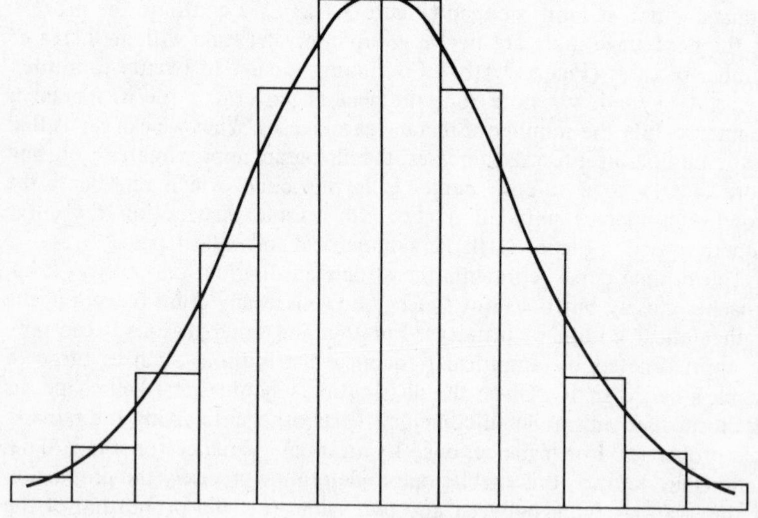

Figure 3.1 (a) and (b) *Histograms and the normal curve.*

distribution and the second its variance. Where the distribution is only approximate, we use instead the symbol $\dot{\sim}$, thus:

$$x \dot{\sim} N(\mu, \sigma^2) \tag{3.10}$$

In determining areas between x-values of a normal distribution, a routine procedure may be used. The method involves changing, or transforming, our x-variable to what is called the *standardised normal variable*, designated by z. This is done by means of the expression

$$z = \frac{x - \mu}{\sigma} \tag{3.11}$$

What we are doing is determining the difference between x and μ in terms of the population standard deviation σ. The variable z is normally distributed with mean 0 and standard deviation 1. Every x-value will have a related z-value. In Figure 3.2 x_1 and z_1 correspond, as do x_2 and z_2; z_1 is a positive number greater than 1 because x_1 is greater than $\mu + \sigma$, but z_2 is negative and between -1 and 0 because x_2 is less than μ but greater than $\mu - \sigma$. The area under the standard normal curve is exactly 1. This can be equated with the total proportion of observations of the x-variable, also 1. In addition, the proportion of observations having values less (or greater) than a particular x-value is the same as the area under the standard normal curve below (or above) the corresponding z-value.

In Table A in the Appendix (pp. 185–6) there are tabulated the areas under the standard normal curve lying below particular z-values. The table is constructed along the same lines as logarithm tables. The z-values in the left-hand column increase by steps of 0.1. Second decimal places in z are obtained by moving across the first set of columns and the third decimal places are derived by adding or subtracting the quantities in the right-most columns. For example, the area below a z-value of 0.852 is 0.8029, i.e. 0.8023 (corresponding to a z-value of 0.85) plus 6 in the fourth decimal place. Similarly, the area below a z-value of -1.645 is 0.0500, i.e. 0.0505 less 5 in the fourth decimal place. As the normal distribution is perfectly symmetrical, the area lying above a z-value of $+1.645$ must also be 0.0500, i.e. the same as the area lying below the z-value, -1.645. Some straightforward examples will help the reader to understand how we use Table A in the Appendix in practice.

Examples. An IQ test has been so constructed (standardised) that for a particular age group the IQ distribution is normal with a mean of 100 and a standard deviation of 15. What proportion of the age group will have an IQ (a) of 105, (b) greater than 135, (c) between 110 and 120 inclusive?

(a) With this problem we immediately appear to hit a snag. The problem is represented in Figure 3.3. The value 105 appears as a vertical line *having no width*. Therefore, we seem unable to obtain a corresponding area from the standard normal curve. The problem is made soluble by regarding 105 (quite properly) as a rounded value, i.e. as representing any value between 104.5 and 105.5. The problem then resolves itself into finding the proportion of observations having values between 104.5 and 105.5. Thus, we need the z-values corresponding to these two limits:

$$z_1 = \frac{104.5 - 100}{15} = 0.300 \qquad z_2 = \frac{105.5 - 100}{15} = 0.367$$

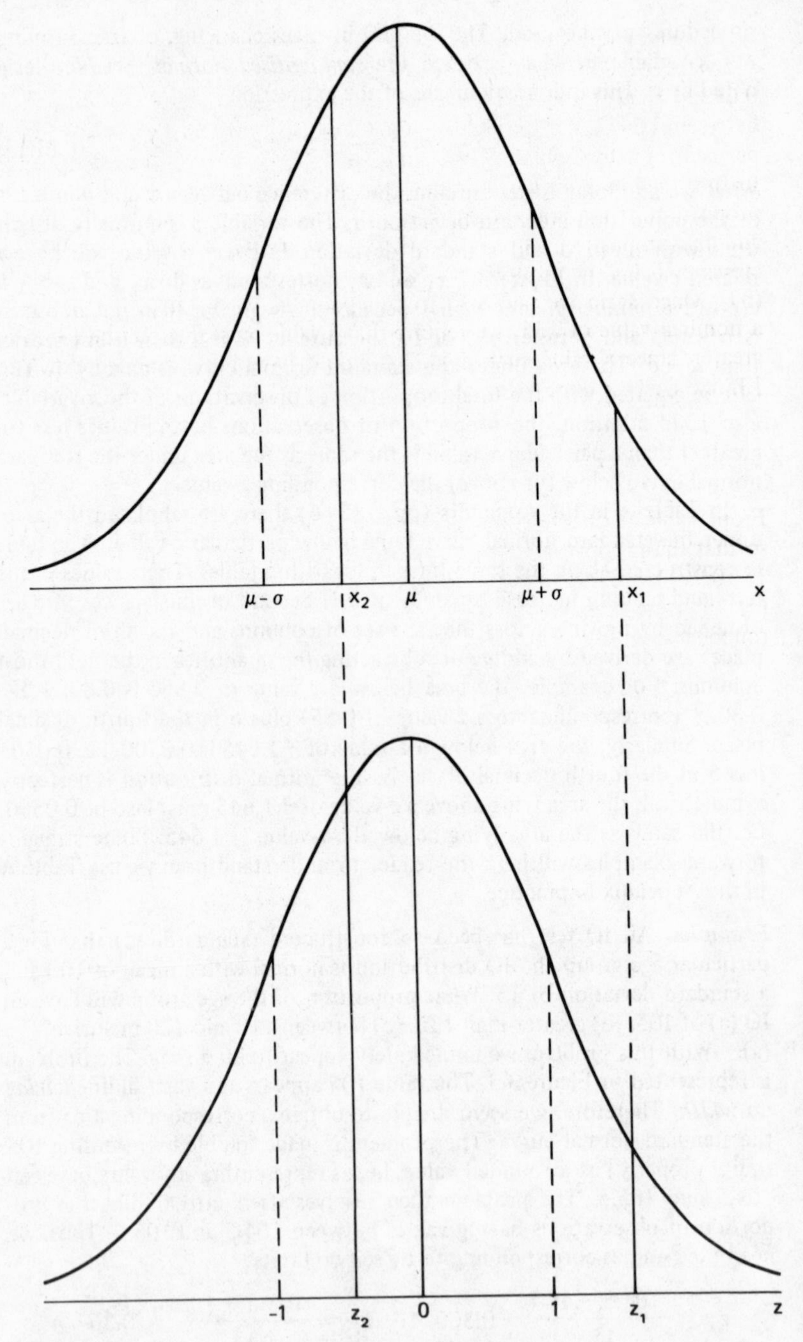

Figure 3.2 *General and standard normal curves.*

If we now refer to Table A in the Appendix we find that the area below a z-value of 0.300 is 0.6179, and the area below a z-value of 0.367 is 0.6432. Hence, the area between these two z-values is the difference between 0.6432 and 0.6179, i.e. 0.0253. So we would say that about 2.5 per cent of our population had an IQ of 105. We must note that given the feature that IQ scores are to be regarded as rounded values, the sum of the proportion of the population with an IQ greater than any value (e.g. 105) and the proportion with an IQ less than that value is not unity, the discrepancy being the proportion *with* that particular value.

(b) Once again we take 135 to be a rounded value. To be greater than a nominal value of 135, we require that a value must in effect be 135.5 or greater, since a value such as 135.4 would have been rounded down to 135. So we base our z-value on 135.5. We have:

$$z = \frac{135.5 - 100}{15} = 2.367$$

From Table A (in the Appendix) the area below this value of z is 0.9911, so that the proportion of our population having an IQ greater than 135 is $1 - 0.9911 = 0.0089$, or 0.89 per cent.

(c) In this case nominal values of 110 and 120 must be considered rounded from 109.5 to 110.5 and 119.5 to 120.5, respectively. A value

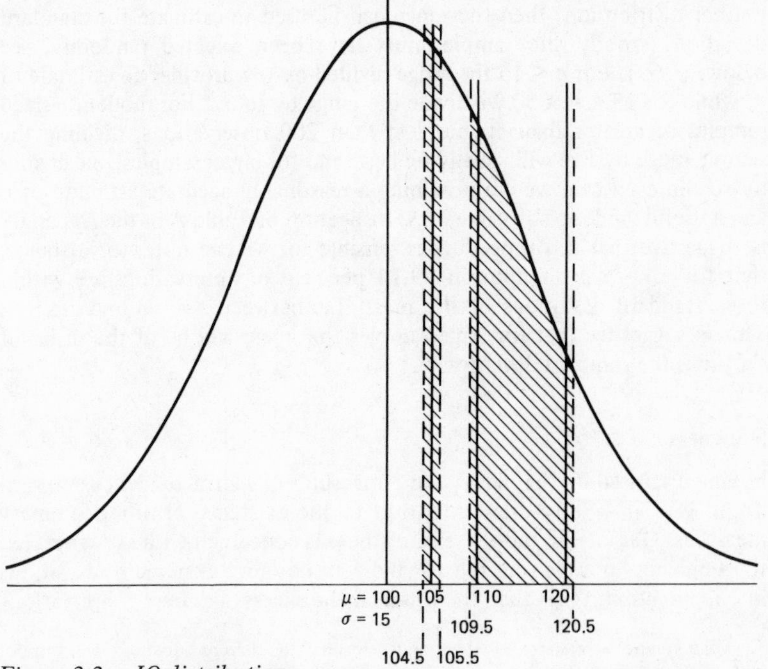

Figure 3.3 *IQ distribution.*

nominally between 110 and 120 inclusive could be between 109.5 and 120.5 before rounding, and these are the limits we use to obtain the z-values (see Figure 3.3):

$$z_1 = \frac{109.5 - 100}{15} = 0.633 \qquad z_2 = \frac{120.5 - 100}{15} = 1.367$$

The areas corresponding to these z-values are 0.7367 and 0.9142. The difference between them of 0.1775 gives the proportion of the population with an IQ between 110 and 120.

One is now in a position to see how important the standard deviation is in the context of the normal distribution. It is used as a standardising value which enables us to transform any normally distributed variable into the standardised normal variable z. Very frequently in what follows we will be using it for this purpose, and herein lies its important place in statistical work.

We noted earlier the importance of obtaining, where possible, advance estimates of measures which we seek to calculate, so as to reduce the chances of error. Having considered the character of the normal distribution, we can add to those remarks. Manifestly, if a distribution approximates to the normal form, the mode will provide an approximate mean value. Also where we have a sample of size n from an approximately normal distribution, then the range can be used to estimate the standard deviation (strictly the sample must have been selected randomly, see below, p. 69). For $n \leqslant 15$ the range divided by $\sqrt{n}$ provides an estimate of σ, while for $15 < n \leqslant 50$ we divide the range by four.* For moderate-sized samples of greater than 50 but less than 200 observations, dividing the sample range by five will usually be best, and for larger samples, one divides by six. In each case we are obtaining a reasonably accurate estimate of σ and a useful guide to the value of S. Inspection of Table A in the Appendix is suggestive that this procedure is sensible for we can note, for instance, that for a normal distribution 99.74 per cent of observations are within three standard deviations of the mean, i.e. between $z = -3$ and $z = +3$. This is suggestive that for large samples the range will be of the order of six times the standard deviation.

Skewness

Having discussed in Chapters 2 and 3 measures of central tendency and variation, we can briefly draw attention to the existence of other summary measures. One of the more useful of these is concerned with *skewness*, i.e. the tendency in a distribution for there to be more extreme observations in one direction rather than the other. If the excess of extreme observations

*The symbol < means 'less than' and ⩽ means 'less than or equal to'. The symbol > is used for 'greater than'.

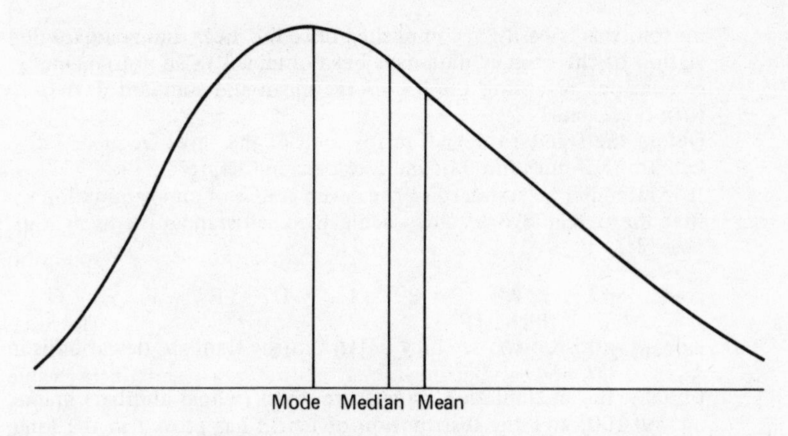

Figure 3.4 *A moderately skewed distribution.*

is in the upward direction the distribution is described as *positively skewed* (this is often true for income distributions; see Table 2.7, p. 16), while if it is in the downward direction the distribution is *negatively skewed*.

Figure 3.4 illustrates a moderately skewed distribution and shows the relative positions of the mean, median and mode. For such distributions the median will generally be about twice as far from the mode as from the mean. A good measure of the skewness of a distribution is given by the formula

$$\text{skewness} = \frac{3\,(\text{mean} - \text{median})}{\text{standard deviation}} \tag{3.12}$$

This is a particular form of the *Pearsonian coefficient of skewness*. It serves to differentiate between positively skewed distributions, where the mean lies above the median, and negatively skewed distributions, where the reverse is true.

Glossary

Range
Quartile deviation
Deciles
Percentiles
Modulus
Mean (or average) deviation
Variance
Standard deviation
Normal (Gaussian) distribution
Pearsonian coefficient of skewness

Exercises

1(a) Obtain the standard deviation for the raw data of question 2 of the Exercises in Chapter 2.

(b) Obtain the mean and standard deviation for these same data *without* including the two values 3,829 and 3,527 (which refer to Hong Kong and Singapore). (If you are using a calculator with an xDEL button, this is easily accomplished once the mean and standard deviation of the original data have been obtained.) Can you draw any conclusions regarding the use of the mean and standard deviation with these data?

2 Obtain the standard deviation for each of the three frequency distributions of question 3 of the Exercises in Chapter 2.

3 It is intended to standardise the seven grades of an examination so that the percentage of entry achieving each standard is as near as possible:

grade	A (highest)	B	C	D	E	F	G (lowest)
percentage	10	15	10	15	20	20	10

Initially the examination papers are given (whole number) marks out of 100, and the distribution of marks has proved in the long term to be approximated by a normal curve with mean 49.5 and standard deviation 15. What do you suggest should be the maximum and the minimum for the marks corresponding to each grade? (Hint: With respect to (say) grade G, inspection of Table A (in the Appendix) shows that 10 per cent (or 0.1) of the area under the standard normal curve falls below a z-value of -1.282. This corresponds to an x (mark) value of $49.5 - 1.282 \times 15 = 30.27$. Hence, marks of 30 and below – with 30 seen as a rounded value with limits of 29.5 and 30.5 – will occur in approximately 10 per cent of cases.)

4 In a large-scale study of the length of time of engagement periods, it was found that this variable was approximately normally distributed and that 2.5 per cent of couples were engaged for less than three months and 15.0 per cent for more than twelve months. Estimate the mean length of time couples were engaged and the standard deviation. (Hint: Inspection of Table A (Appendix) shows that 2.5 per cent of the area under the standard normal curve falls below a z-value of -1.960, hence $\mu - 1.960\sigma = 3$. A second equation may be similarly obtained.)

Chapter 4

Probability

Definitions of Probability

From birth to death, the consequences of our actions are governed by probabilities. When we are born, there is a probability that we will live to a particular age. This probability will increase as we approach that age. When we walk or drive along a road, there is a probability that we will be involved in an accident. If we work in industry, there is a probability that we may be made redundant next week. There is a probability that on a particular day it will rain. What, then, do we mean by 'probability'? Whatever it is, its importance, both in statistical work and our everyday lives, cannot be overemphasised.

Defining probability may appear easy but in reality is very difficult. There are two commonly used definitions and both are quite easy to understand, but both have quite serious defects, however useful they may turn out to be in practice. As the notion of probability applies only to situations where there is uncertainty, these are such that there is always more than one possible outcome. A simple example consists of throwing a six-sided die. It must land in one of six ways. If the die is a fair one, there will be no way of knowing in advance which of the six sides, numbered 1 to 6, will turn up. All we can say is that, as long as the die is perfect, then each side will have exactly the same chance of turning up as any of the others. Intuitively we would say that there was a one-in-six chance of, for instance, a 6 turning up. This provides us with the first definition of probability:

> The *probability of an event* E *occurring* is the total number of equally likely outcomes divided into the number of those outcomes which satisfy the requirement that E occurs.

If we let the total number of equally likely outcomes be N and the number of these where E happens is h, and if we represent the probability of E occurring by $Pr(E)$, then

$$Pr(E) = \frac{h}{N} \tag{4.1}$$

Because h can only take values between 0 and N, the probability of an event occurring must be a number between 0 and 1. With our fair die, if we require the probability of a 6, $Pr(6)$, the total number of equally likely outcomes is six and of these only one represents the number 6, so with $N = 6$ and $h = 1$, $Pr(6) = \frac{1}{6}$. By similar reasoning, if we require the probability that an even number is obtained, we have $N = 6$ and $h = 3$, so Pr (even number) $= \frac{3}{6} = \frac{1}{2}$.

Note that there is an explicit assumption about the physical nature of the object used. The die must be bona fide, i.e. it must be perfectly symmetrical and homogeneous. But in practice we may not know if the assumption is true. We might provisionally have to rely on the maker's word that it is so. The problem with this definition lies in the phrase 'equally likely outcomes'. We may believe we know what is meant by it, but if we analyse it further, we see that it is synonymous with 'equally probable outcomes'. So we are, in effect, using the very term that we are trying to define. It is not very satisfactory simply to provide a circular definition. However, if our assumption about equally likely outcomes is true, this definition is very useful in practice and is a basis upon which probability theory may be built.

However, what if our assumption of equally likely outcomes is incorrect? How can we determine the probabilities involved with a biased die? In practice, we cannot do so exactly. All we can do is to obtain estimates of the probabilities by empirical or experimental methods. We might throw the die a number of times ('repeated trials' in statistical terminology) and note in how many trials each particular event occurred. If for each event we then divided the number of throws n into the number of times the event happened m, we would obtain a set of relative frequencies which would be estimates of the true probabilities. The more times we threw the die, the more confidence we would have in our estimates. But however many times the experiment was repeated, we could never be sure that our estimates exactly equalled the true probabilities. This leads us to the alternative definition of probability based on relative frequency:

$$\textit{Probability of } \mathrm{E} \textit{ occurring}, Pr(E) = \lim_{n \to \infty} m/n \qquad (4.2)$$

where this latter quantity means the value of m/n as n becomes indefinitely large. What this essentially mathematical definition implies is that as the number of experiments (or trials) increases, so the relative frequency tends to get nearer to the true probability, though equality is not guaranteed until the number of trials is infinite, which in reality means never.

Suppose we threw the die a large number of times, and after every twenty throws we noted how many 6s had been observed up to that point. Our results might look something like those in Table 4.1. It would be useful for the reader to compile a similar table for himself.

Figure 4.1 illustrates these results graphically. We can see that the relative frequency, though changing fairly rapidly at first, appears to be settling down around a value in the vicinity of 0.16. That we would expect

Table 4.1 *Outcomes of Dice Throwing*

Number of throws n	*Number of sixes* m	*Relative frequency* m/n	n	m	m/n
20	5	0.250	220	36	0.164
40	8	0.200	240	39	0.163
60	10	0.167	260	41	0.158
80	12	0.150	280	42	0.150
100	15	0.150	300	45	0.150
120	20	0.167	320	47	0.147
140	24	0.171	340	52	0.153
160	26	0.163	360	56	0.156
180	27	0.150	380	58	0.153
200	30	0.150	400	63	0.158

the fluctuations to become more and more damped as the number of trials increases, is suggested intuitively by the following consideration. As the number of trials increases, the value of the relative frequency is less substantially affected by particular unlikely events. For instance, had we obtained ten 6s in (say) the twenty trials from the 101st to the 120th (itself a highly unlikely event), the relative frequency would have risen from 0.150 to 0.208, whereas obtaining ten 6s in the twenty trials from the 301st to the 320th would only have raised the relative frequency from 0.150 to 0.172.

In our experiment it might seem that there is some evidence to suggest that the die may not be fair, since the probability of getting a 6 when throwing a perfect die is 1:6 or 0.167. After 400 trials, we might have

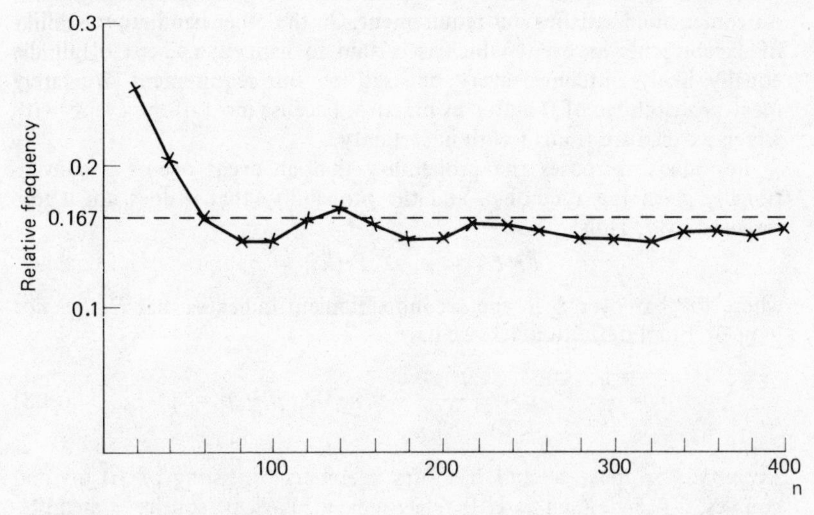

Figure 4.1 *Dice throws.*

thought the relative frequency would have settled down to values around this, but in fact it fails to reach this value from the 160th trial onwards (for the n shown in Table 4.1). However, our results are by no means conclusive and we would have to continue our experiment for over 6,500 trials (!) before we might reasonably conclude that the die was biased, assuming that the relative frequency was still as far from the 0.167 value at that stage.

In real-life situations it is often impossible to calculate probabilities using definition 4.1, and one is forced to use estimates based on relative frequencies as suggested by definition 4.2. A simple example will serve as an illustration. What is the probability that a mother-to-be will give birth to a male (or female) child? All we can say is that in the past 51.4 per cent of all births have been male and 48.6 per cent female, and these figures have varied little from generation to generation. So, for an individual birth (assuming that no medical tests are carried out which reveal the baby's sex), the probability of the child being male can be considered to be 0.514. This, of course, is still an estimate of the probability, but under the circumstances it must be a very accurate one.

As far as the building up of a theory of probability is concerned the relative frequency definition presents problems, since in practice all we have are estimates, and we generally do not know by how much our estimates differ from the true probabilities. So the first definition, 4.1, is, despite its faults, used here as a basis for probability theory.

Calculating Probabilities

In determining probabilities, we must note that a probability of 0 represents an event which cannot happen, i.e. out of all the equally likely outcomes none satisfies our requirement. On the other hand, a probability of 1 represents an event which is certain to happen, i.e. out of all the equally likely outcomes every one satisfies our requirement. We rarely meet probabilities of 0 and 1 in practice, because most of the events with which we deal are fraught with uncertainty.

For many purposes the probability that an event occurs is conventionally given the symbol p, and the probability that it does not is represented by q. Thus:

$$Pr(E) = p \qquad Pr(\bar{E}) = q$$

where the bar over E in the second statement indicates that E does *not* happen. From definition 4.1, we have

$$p = \frac{h}{N} \qquad q = \frac{N-h}{N} \qquad \text{so that } p + q = 1 \tag{4.3}$$

Examples. Suppose a club has fifty members consisting of 10 married couples, 20 single men and 10 single women. Each person has a membership card bearing a number from 1 to 50. We choose a member of the club by a method which ensures that each person is equally likely to be chosen. There are a number of ways of doing this but the method we adopt is to 'draw lots', i.e. select one from fifty slips which have been marked with numbers from 1 to 50 and shuffled in a container. The required individual has the selected membership number. Let us determine some probabilities concerning the sex and marital status of the chosen person.

First, we may consider (a) the issue of the selected person's sex. Let E_1 be the event that a man is chosen and p be the associated probability. There are altogether 30 men (10 married and 20 single) among the fifty club members. So

(a) $$Pr(E_1) = p = 30/50 = 0.6$$

There are also 20 women club members (10 married and 10 single), so if E_2 represents the event of selecting a woman,

$$Pr(E_2) = 20/50 = 0.4$$

We can see that $Pr(\bar{E}_2) = Pr(E_1) = p$, because the event that a woman is not selected corresponds exactly to the event that a man is selected. As an illustration of result 4.3, we have

$$Pr(E_1) + Pr(E_2) = Pr(E_1) + Pr(\bar{E}_1) = 0.6 + 0.4 = 1$$

We can next ask (b) what is the probability of selecting from the club members either a single man or a married woman, and (c) what is the probability of selecting either a single woman or a married person? From definition 4.1:

(b) $$Pr(E) = Pr\text{ (single man or married woman)} = \frac{20 + 10}{50} = 0.6$$

(c) $$Pr(E) = Pr\text{ (single woman or married person)} = \frac{10 + 20}{50} = 0.6$$

Note that in these cases we can add the individual probabilities together. For instance, in (b)

$$Pr(E) = Pr\text{ (single man)} + Pr\text{ (married woman)} = \frac{20}{50} + \frac{10}{50} = \frac{30}{50}$$

Let us next introduce a slightly more complex notion. Given two events E_1 and E_2, by the *conditional probability* of E_2 given E_1 – denoted by $Pr(E_2|E_1)$ – we mean the probability that E_2 occurs given that E_1 has occurred. Returning to the selection of club members, we can this time consider drawing a numbered slip from the container but then drawing a second slip without returning the first one, so that we have two numbers representing two people (a simple example of what is termed *sampling without replacement*). If $Pr(E_1)$ is the probability of choosing a man on the first draw and $Pr(E_2)$ that of choosing a man on the second draw, what is $Pr(E_2|E_1)$, i.e. the probability of choosing a man on the second draw given that one has been chosen on the first draw? Now $Pr(E_1) = Pr(E_2) = 30/50$, but if a man is selected on the first draw, is the probability that one will be selected on the second draw still 30:50? We started off with fifty members of whom 30 were men. If the first selection is a man, then the second must be made from 49 members, only 29 of whom are men. So the probability that a man is selected on the second draw is then 29:49, and this is the required conditional probability $Pr(E_2|E_1)$. The fact that a man was selected on the first draw has changed the nature of the population from which the second selection is made.

We have the following definition. If

$$Pr(E_2|E_1) = Pr(E_2)$$

then the two events E_1 and E_2 are said to be *independent events*. It also follows that $Pr(E_1|E_2) = Pr(E_1)$. This definition is particularly important, since most of the events with which we deal in the remainder of this book are of this type. What it says is that the fact that E_1 has happened changes nothing as far as the probability of E_2 is concerned, and vice versa. For example, consider tossing a coin. Suppose that E_1, E_2, E_3, etc. are the events that we get a Head on the first, second, third and subsequent tosses. Assuming that it is a fair coin, then for each throw:

$$Pr(\text{Head}) = Pr(\text{Tail}) = \tfrac{1}{2}$$

$$\text{and } Pr(E_1) = Pr(E_2) = Pr(E_3) = \ldots = \tfrac{1}{2}$$

On any particular toss, the probability of getting a Head is the same. Suppose the first toss turns up a Head. What is the probability that the second will also be a Head, i.e. what is $Pr(E_2|E_1)$? It is still the same irrespective of the result of the first toss!

In this connection confusion may occur between two different aspects of probability. The first is where one is considering, for instance, the probability of throwing a coin three times and obtaining three Heads which is 1:8. We can see this because with H representing Heads and T Tails, there are eight possible equally likely outcomes corresponding to *HHH*, *HHT*, *HTH*, *THH*, *HTT*, *THT*, *TTH*, *TTT*. The other is the probability of tossing a coin for a third time and obtaining a Head, having already obtained Heads on the first two tosses, which is 1:2. The idea that, if Heads have turned up several times in succession, then the probability of a Head on the next toss is less than 1:2 seems very widespread. It is the result of muddled thinking and seems to have lost naïve gamblers more money than any other idea. If the coin is fair, all the events have the same probability and they are independent. Of course, if the coin is biased – possible if there has been a run of Heads – the probability of a further Head may be greater than 1:2. But following such a run, there is no rational way of concluding that the probability of another Head turning up is less than 1:2.

Let us now turn explicitly to consider the probabilities of compound events. For example, if we throw two dice, what is the probability that we will obtain two 6s? We can still work this out from the basic definition 4.1. The total number of equally likely outcomes is 36, made up thus:

Die 1	Die 2	Die 1	Die 2	Die 1	Die 2	Die 1	Die 2	Die 1	Die 2	Die 1	Die 2
1	1	2	1	3	1	4	1	5	1	6	1
1	2	2	2	3	2	4	2	5	2	6	2
1	3	2	3	3	3	4	3	5	3	6	3
1	4	2	4	3	4	4	4	5	4	6	4
1	5	2	5	3	5	4	5	5	5	6	5
1	6	2	6	3	6	4	6	5	6	6	6

Note that, as far as this specification of outcomes is concerned, order is important, e.g. the outcome 1 on die 1, 2 on die 2 is different from the outcome 2 on die 1, 1 on die 2 (even though with a single throw of identical dice the two outcomes could not be distinguished).

We see that only one of the 36 outcomes satisfies our requirement that there be two 6s, i.e. 6 on die 1, 6 on die 2. So the probability of two 6s turning up is 1:36 which is the same as $1/6 \times 1/6$. Put this latter way, we see that the probability of two 6s is the probability of the first die giving a 6 (1:6) multiplied by the probability of the second die giving a 6 (1:6). The two events, E_1 being the probability of 6 on the first die, E_2 being the probability of 6 on the second die, are independent, as the result on either die has no influence on the probabilities for the other. We shall adopt the following notation: E_1E_2 will represent the event that *both* E_1 and E_2 occur. For independent events, we have:

$$Pr(E_1E_2) = Pr(E_1) \times Pr(E_2) \tag{4.4}$$

More generally, for events which are not necessarily independent, we have:

$$Pr(E_1E_2) = Pr(E_1) \times Pr(E_2|E_1) = Pr(E_2) \times Pr(E_1|E_2) \tag{4.5}$$

The understanding of the latter expression is that, if we have any two events, the probability of both occurring is the probability of getting one of these events times the probability of getting the other given that the first event has occurred.

The general result 4.5 is not difficult to establish. Suppose that with respect to these various events there is a total of N equally likely outcomes and that h_1 involve the occurrence of E_1, h_2 the occurrence of E_2, and h_{12} the occurrence of both E_1 and E_2. From definition 4.1, we have:

$$Pr(E_1) = \frac{h_1}{N}; \quad Pr(E_2) = \frac{h_2}{N}, \quad Pr(E_1E_2) = \frac{h_{12}}{N}$$

On the other hand, the conditional probability $Pr(E_2|E_1)$ represents the fraction of outcomes in which E_1 occurs where E_2 also occurs; therefore, assuming h_1 is non-zero, we have:

$$Pr(E_2|E_1) = \frac{h_{12}}{h_1}$$

Rewriting we obtain:

$$Pr(E_2|E_1) = \frac{h_{12}/N}{h_1/N} = \frac{Pr(E_1E_2)}{Pr(E_1)}$$

By similar reasoning

$$Pr(E_1|E_2) = \frac{Pr(E_1E_2)}{Pr(E_2)}$$

and we, therefore, obtain result 4.5.

Example. Returning to selection from the fifty club members, we might now ask the question: What is the probability that if we select two members without replacement, they will both be men?

Let E_1 be the event that the first member is a man, and E_2 the event that the second member is a man:

$$Pr(E_1) = 30/50 \qquad Pr(E_2|E_1) = 29/49$$

and from expression 4.5, $Pr(E_1E_2) = 30/50 \times 29/49$.

An alternative to the method of selection used in this example is *sampling with replacement*. In choosing two club members by this latter method, we would select the first person by drawing a numbered slip from the container and then subsequently choose a second person, but only *after* returning the first slip. In that case there would be fifty club members including 30 men to choose from on each draw, so the two events of choosing a man on each draw would be independent. It follows that the probability of selecting with replacement two members who are male is $(30/50) \times (30/50)$ as given by formula 4.4. Of the two methods, sampling without replacement is by far the more commonly used in practice and we can see why from our example. If the first member's number is selected and replaced, then it has another chance of being chosen on the second draw. Hence, our sample of two individuals may consist only of one (selected twice)! This is a situation which we generally seek to avoid, since for most purposes we require our sample to consist of a certain number of essentially different cases (as when selecting a committee).

The relationships for obtaining the probabilities that two events both occur can be extended to any number of events. For three events all of which are independent of each other:

$$Pr(E_1E_2E_3) = Pr(E_1) \times Pr(E_2) \times Pr(E_3) \qquad (4.6)$$

This gives rise to the expression 'the *product rule*' for independent events, since all we have to do is to multiply the individual probabilities together. Where the events are dependent, the corresponding expression is

$$Pr(E_1E_2E_3) = Pr(E_1) \times Pr(E_2|E_1) \times Pr(E_3|E_1E_2) \qquad (4.7)$$

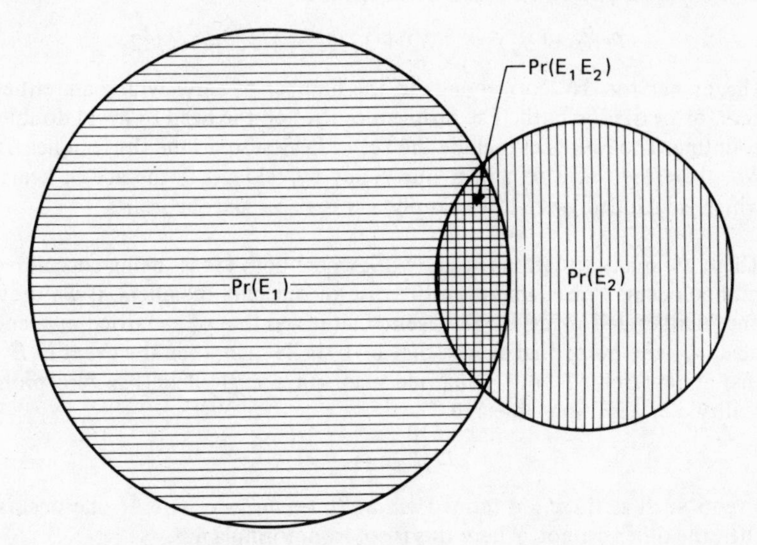

Figure 4.2 *Representation of* $\Pr(E_1 \cup E_2)$.

The term $Pr(E_3|E_1E_2)$ means the probability of E_3 occurring given that both E_1 and E_2 have already occurred.

We now define another type of event – the event that either E_1 or E_2 or both occur. The notation we shall use for this is $E_1 \cup E_2$. The probability of such an event may be obtained from the formula:

$$Pr(E_1 \cup E_2) = Pr(E_1) + Pr(E_2) - Pr(E_1E_2) \qquad (4.8)$$

The $-Pr(E_1E_2)$ term may initially be confusing, since there is a tendency to expect a positive sign. The reason for it becomes clearer on inspection of Figure 4.2. There the probabilities of E_1 and E_2 have been represented by two circular areas (but the precise shape of the areas is of no significance). The probability of both E_1 and E_2 occurring, i.e. $Pr(E_1E_2)$, is represented by the area formed where the two circles intersect. On the other hand, the probability of the event $E_1 \cup E_2$ is represented by the total area bounded by the two circles. It can be seen that this particular area may be considered as formed by the area corresponding to $Pr(E_1)$ taken together with that corresponding to $Pr(E_2)$ but without the area $Pr(E_1E_2)$. If the latter area were not subtracted, it would be counted twice in the addition of the areas $Pr(E_1)$ and $Pr(E_2)$.

Example. Suppose after shuffling we choose a card from a pack, what is the probability that this card is either an ace or a heart, or both?

Let E_1 be the event that the card is an ace, then $Pr(E_1) = 4/52$. Let E_2 be the event that the card is a heart, then $Pr(E_2) = 13/52$. E_1E_2 is the event that the card is both an ace and a heart, i.e. it is the ace of hearts. $Pr(E_1E_2) = 1/52$ and from formula 4.8 we have:

$$Pr(E_1 \cup E_2) = 4/52 + 13/52 - 1/52 = 16/52$$

The numerator, 16, corresponds to the number of cards which are either aces, or hearts, or both. The problem centres on the need to avoid double-counting. The four aces include the ace of hearts; so do the thirteen hearts. We, therefore, have to adjust our count by taking off the ace of hearts which is included twice if we simply add the aces and the hearts.

There are many situations in our work where both events being considered cannot occur. For example, in the case of our club members, if we drew one number and made E_1 the event that it was that of a married man and made E_2 the event that it was that of a single man, then the event E_1E_2, that it represented both a married man and a single man, has zero probability, i.e. it is impossible and

$$Pr(E_1E_2) = 0 \tag{4.9}$$

Events such as these are termed *mutually exclusive events*. If one occurs, then the other cannot. Where this is so, from formula 4.8,

$$Pr(E_1 \cup E_2) = Pr(E_1) + Pr(E_2) \tag{4.10}$$

Again, this result can be extended to any number of events. For three such events:

$$Pr(E_1 \cup E_2 \cup E_3) = Pr(E_1) + Pr(E_2) + Pr(E_3) \tag{4.11}$$

which gives rise to what is termed 'the *addition rule*' for mutually exclusive events.

Example. Again using a pack of cards, if we draw one card and let E_1 be the event that it is a king, E_2 the event that it is a queen and E_3 the event that it is a knave, then since we have three mutually exclusive events, the probability of a court card is given by,

$$Pr(E_1 \cup E_2 \cup E_3) = 4/52 + 4/52 + 4/52 = 12/52$$

Combinations and Probability

Suppose that we wished to solve the following problem. Our fifty club members decide to form a committee of three. No one is very keen to stand for election. In fact, the married couples opt out entirely, leaving only the single men (20) and the single women (10). It is decided to select the three committee members by placing the 30 available members' numbers into a container, shaking it and successively drawing out three. What is the probability that there will be two men and one woman on the committee, i.e. that there will be proportional representation? We can solve this problem in the following way.

Our requirement that there be two men and one woman will be satisfied, if the numbers drawn represent one of the following sequences –

MMW or *MWM* or *WMM* (M = man, W = woman). Each of these three outcomes will have exactly the same probability, for in determining the magnitude of any one the same numbers must be present each time in numerator and denominator, so let us find the probability of the sequence *MMW* occurring. There will be 30 numbers which can be drawn to obtain the first committee member, 29 for the second and 28 for the third. Altogether there will be $30 \times 29 \times 28$ different equally likely outcomes. These form the denominator for the required probability. We now have to determine how many of these outcomes satisfy the requirement that there are two men followed by one woman. On the first draw there are 20 possible numbers which are those of men, while on the second draw there are 19. So two men can be selected in 20×19 ways. The one woman can then be selected in 10 different ways. In total there are $20 \times 19 \times 10$ different ways of selecting three members such that we obtain two men followed by one woman. The probability of this event by definition 4.1 is

$$\frac{20 \times 19 \times 10}{30 \times 29 \times 28}$$

This result could also have been obtained by considering the conditional probabilities of the three events – man, followed by man, followed by woman. The probability that the first person selected is a man is 20:30. The probability that the second person selected is a man given that the first was a man is 19:29. The probability that the third person selected is a woman given that the first two were men is 10:28. Thus, from expression 4.7, the probability of the required sequence is:

$$Pr(MMW) = Pr(M) \times Pr(M|M) \times Pr(W|MM) = \frac{20}{30} \times \frac{19}{29} \times \frac{10}{28}$$

We now have the probability of one of the three possible sequences. The three are examples of mutually exclusive events. We require the probability that one or other occurs, i.e. $Pr(MMW \cup MWM \cup WMM)$. In this case, since they each have the same probability, we can multiply the probability of *MMW* by three. The final result is

$$Pr(\text{2 men and 1 woman}) = 3 \times \frac{20}{30} \times \frac{19}{29} \times \frac{10}{28}$$

The above methods of obtaining probabilities can be rather time-consuming. There are more elegant ways of achieving the same results. But as a preliminary, we have to define some new symbols. The first is $n!$, stated 'n-factorial'. By definition,

$$n! = n(n-1)(n-2)(n-3)\ldots 3 \times 2 \times 1$$

where n is a whole number greater than or equal to one. In addition $0!$ is defined as being equal to 1. The second symbolic representation concerns processes of selection. The number of different ways in which r items can be selected from n items is known as the number of combinations of r from n. It was originally given the symbol ${}^nC_r$, but this has now generally been changed to $\binom{n}{r}$. It can be calculated thus:

$$\binom{n}{r} = \frac{n!}{r!(n-r)!} \quad (\text{for } r \leqslant n) \qquad (4.12)$$

Verbally, the number of ways in which r items can be selected from n is n-factorial divided by the product of r-factorial and $(n-r)$-factorial. Because of the definition $0! = 1$, we have the results:

$$\binom{n}{n} = 1 \quad \text{and} \quad \binom{n}{0} = 1$$

We also note that

$$\binom{n}{r} = \frac{n!}{r!(n-r)!} = \binom{n}{n-r}$$

that is, the number of selections of r items from n is exactly the same as the number of selections of $(n-r)$ items from n. This result can sometimes be of use and one can observe, for instance, that

$$\binom{20}{16} = \binom{20}{4} = \frac{20 \times 19 \times 18 \times 17}{4 \times 3 \times 2 \times 1}$$

Examples. (a) If we seek to determine the number of ways in which two items can be selected from six:

$$\binom{6}{2} = \frac{6!}{2!\,4!} = \frac{6 \times 5 \times \cancel{4} \times \cancel{3} \times \cancel{2} \times \cancel{1}}{2 \times 1 \times \cancel{4} \times \cancel{3} \times \cancel{2} \times \cancel{1}} = 15$$

(b) Again, for the number of selections of two from seven items:

$$\binom{7}{2} = \frac{7!}{2!\,5!} = \frac{7 \times 6 \times \cancel{5} \times \cancel{4} \times \cancel{3} \times \cancel{2} \times \cancel{1}}{2 \times 1 \times \cancel{5} \times \cancel{4} \times \cancel{3} \times \cancel{2} \times \cancel{1}} = 21$$

Note that the last $(n-r)$ digits in the numerator always cancel with the $(n-r)$-factorial in the denominator, leaving r digits in the numerator, and r digits from the r-factorial in the denominator.

We can easily see that the last two results are correct by imagining six people labelled A, B, C, D, E and F (and a seventh G) from whom we have to select two. How many different combinations are there? We can list them:

AB	*BC*	*CD*	*DE*	*EF*	(*FG*)
AC	*BD*	*CE*	*DF*	(*EG*)	
AD	*BE*	*CF*	(*DG*)		
AE	*BF*	(*CG*)			
AF	(*BG*)				
(*AG*)					

As can be seen with six people there are 15 possibilities. When person G is introduced there are 6 further possibilities, i.e. 21 in all. The use of combinations simplifies problems where we wish to know how many different ways there are of selecting a certain number of items from a larger number.

To return to our earlier stated problem of selecting a committee of 3 members such that 2 are men and 1 is a woman, then using the formulae for combinations:

$$Pr\,(2 \text{ men and } 1 \text{ woman}) = \frac{\binom{20}{2}\binom{10}{1}}{\binom{30}{3}}$$

In words, the total number of equally likely ways of selecting 3 people from 30 is the number of combinations of 3 from 30, i.e. $\binom{30}{3}$; of these the number that satisfies the requirement that 2 are men and 1 is a woman is obtained by multiplying the number of ways of selecting 2 men from 20, i.e. $\binom{20}{2}$, by the number of ways of selecting 1 woman from 10, i.e. $\binom{10}{1}$. Expanding the above expression we obtain

$$Pr\,(2 \text{ men and } 1 \text{ woman}) = \frac{\dfrac{20 \times 19}{2 \times 1} \times \dfrac{10}{1}}{\dfrac{30 \times 29 \times 28}{3 \times 2 \times 1}} = \frac{3 \times 20 \times 19 \times 10}{30 \times 29 \times 28}$$

the same result as we had previously.

We can next usefully consider the other possible combinations of men and women in our committee of three. We could have all men and no woman, one man and two women or all women. What are the probabilities of these events? It is convenient to construct a table where r denotes the number of men (Table 4.2). It is very easy to see the correspondence between the first (no man) and the last (no woman) probabilities and the

Table 4.2 *Sex Composition of a Committee*

r	$Pr(r)$
0	$\binom{10}{3} \Big/ \binom{30}{3} = 0.0296$
1	$\binom{20}{1}\binom{10}{2} \Big/ \binom{30}{3} = 0.2217$
2	$\binom{20}{2}\binom{10}{1} \Big/ \binom{30}{3} = 0.4680$
3	$\binom{20}{3} \Big/ \binom{30}{3} = 0.2808$

results we would have obtained from first principles. From first principles,

$$Pr(0 \text{ men}) = Pr(\text{all women}) = \frac{10}{30} \times \frac{9}{29} \times \frac{8}{28}$$

and

$$Pr(\text{all men}) = \frac{20}{30} \times \frac{19}{29} \times \frac{18}{28}$$

Whereas using combinations,

$$Pr(0 \text{ men}) = \frac{\binom{10}{3}}{\binom{30}{3}} = \frac{10 \times 9 \times 8}{30 \times 29 \times 28}$$

and

$$Pr(\text{all men}) = \frac{\binom{20}{3}}{\binom{30}{3}} = \frac{20 \times 19 \times 18}{30 \times 29 \times 28}$$

In the table all possible events are included – the list is exhaustive – so one or other must happen. They are also mutually exclusive events (if there are two men there cannot be no man, etc.) so the addition rule applies. The total of the probabilities should therefore equal one, and we can confirm that this is so. We have here what is termed a *probability distribution*. All possible values have been included and the sum of their probabilities is one. These are the requirements for probability distributions for discrete variables. Such distributions are very important indeed in statistics, and we are now in a position to introduce one such distribution – the binomial.

The Binomial Distribution

Visualise a situation where five pregnant women are in hospital together. All are due to have babies at much the same time. It is known that no birth will be multiple. What is the probability that there will be 3 boys and 2 girls born to the 5 women? For calculation purposes, we shall take the probability of a male birth to be 0.52, and that of a female birth to be 0.48. Labelling the mothers-to-be A, B, C, D and E, let us, initially, consider the event that A, B and C give birth to boys and D and E give birth to girls. The associated probabilities are:

	A	B	C	D	E
Probabilities	0.52	0.52	0.52	0.48	0.48

These five events are independent of each other so that, as we require the probability that they all occur, the multiplication rule will apply. The probability that the births occur in this way is thus:

$$0.52 \times 0.52 \times 0.52 \times 0.48 \times 0.48 = (0.52)^3(0.48)^2$$

However, this is *not* the probability that 3 boys and 2 girls are born – it is only the probability that this arbitrarily chosen arrangement of births will occur. There are other arrangements which would satisfy the requirement that there be 3 boys and 2 girls. Boys might be born to A, C and E, and girls to B and D; or boys to C, D and E with girls to A and B, and so on. Each of these will have the same probability of occurring, i.e. $(0.52)^3$ $(0.48)^2$. Just how many different arrangements are there? This is a simple example of the use of combinations. The number of possible arrangements is exactly the same as the number of ways of selecting 3 from 5, i.e. $\binom{5}{3} = 10$. There are thus ten arrangements, all mutually exclusive, and each with probability $(0.52)^3(0.48)^2$. So the probability that one or other of them will happen, which is the probability of 3 boys and 2 girls being born, will be $\binom{5}{3}(0.52)^3(0.48)^2$.

In similar fashion we can obtain the probabilities for the other possible sex combinations and build them into a table (Table 4.3).. Once again, the probabilities of the first and last values, 0 and 5, could have been obtained easily from first principles. Note the pattern in the probabilities. The combinations term is always $\binom{5}{r}$. The probability of a male birth is 0.52, and its power in the probability column is always equal to r; 0.48 is the probability of a female birth, and its power is always equal to $5 - r$.

This type of pattern is always present where n events are being considered, each of which may occur with probability p, or fail to occur with probability $q = 1 - p$. The probability of r of the n occurring is given by the expression:

Table 4.3 *Male Births to Five Mothers*

Number of male births r	Probability $Pr(r)$
0	$(0.48)^5 = 0.0255$
1	$\binom{5}{1}(0.52)(0.48)^4 = 0.1380$
2	$\binom{5}{2}(0.52)^2(0.48)^3 = 0.2990$
3	$\binom{5}{3}(0.52)^3(0.48)^2 = 0.3240$
4	$\binom{5}{4}(0.52)^4(0.48) = 0.1755$
5	$(0.52)^5 = 0.0380$

$$Pr(r) = \binom{n}{r} p^r q^{n-r} \tag{4.13}$$

This is known as the *general term of the binomial distribution*. It can be shown that all terms of this type, with values of r from 0 to n, are given by the following expansion:

$$(q+p)^n = q^n + \binom{n}{1} pq^{n-1} + \binom{n}{2} p^2 q^{n-2} + \binom{n}{3} p^3 q^{n-3} + \ldots + p^n \tag{4.14}$$

The terms of this binomial expansion correspond to

$$Pr(0) \quad Pr(1) \quad Pr(2) \quad Pr(3) \quad \ldots \quad Pr(n)$$

and because $q + p = 1$, and 1 to any power is still 1, the sum of these probabilities is 1, which confirms that the binomial satisfies the condition for a probability distribution. The expression 'binomial' refers to the fact that the expansion is based on the sum of two elements. The binomial is undoubtedly the most important discrete distribution in statistics. The reason why r is used to denote the variable here is purely conventional. (Most publications use x to denote variables in frequency distributions, see Chapter 2, and r for variables in discrete probability distributions.)

We may define the mean and variance of a probability distribution in a way which is essentially analogous to that used with frequency distributions. From expressions 2.3 and 3.8, we can write down for frequency distributions

$$\text{mean} = \frac{\Sigma f_i x_i}{\Sigma f_i} \qquad \text{var.} = \frac{\Sigma f_i x_i^2}{\Sigma f_i} - \left[\frac{\Sigma f_i x_i}{\Sigma f_i}\right]^2$$

with the summation being over k categories. Since Σf_i is simply the total number of observations n, the above can be modified to read

$$\text{mean} = \sum x_i\left(\frac{f_i}{n}\right) \qquad \text{var.} = \sum x_i^2\left(\frac{f_i}{n}\right) - \left[\sum x_i\left(\frac{f_i}{n}\right)\right]^2$$

If we now replace the x_i by the r-variable, and also refer back to the definition of probability based on relative frequency given in expression 4.2 to note that as n becomes indefinitely large the value of f_i/n is $Pr(r)$, we obtain

$$\text{mean} = \sum_{r=0}^{n} rPr(r) \qquad \text{var.} = \sum_{r=0}^{n} r^2 Pr(r) - \left[\sum_{r=0}^{n} rPr(r)\right]^2$$

$$= \sum_{r=0}^{n} r^2 Pr(r) - (\text{mean})^2$$

Although these latter expressions constitute definitions, they become more meaningful once the parallelism with frequency distributions is realised. (One must observe that whereas with frequency distributions the summations are taken from $i = 1$ to $i = k$, the corresponding sums for probability distributions are from $r = 0$ to $r = n$.)

So in order to find the mean for any probability distribution, all we have to do is to sum the products of the r-values and their corresponding probabilities. If we do this for the binomial, we obtain

$$\sum_{r=0}^{n} rPr(r) = 0 \times q^n + 1 \times \binom{n}{1} pq^{n-1} + 2 \times \binom{n}{2} p^2 q^{n-2}$$

$$+ 3 \times \binom{n}{3} p^3 q^{n-3} + \ldots + np^n$$

$$= 0 + npq^{n-1} + 2 \times \frac{n(n-1)}{2 \times 1} p^2 q^{n-2}$$

$$+ 3 \times \frac{n(n-1)(n-2)}{3 \times 2 \times 1} p^3 q^{n-3} + \ldots + np^n$$

$$= npq^{n-1} + n(n-1)p^2 q^{n-2}$$

$$+ \frac{n(n-1)(n-2)}{2 \times 1} p^3 q^{n-3} + \ldots + np^n$$

It can be observed that each term contains both n and at least one p and we can take these outside the expression, giving

$$np\left[q^{n-1} + (n-1)pq^{n-2} + \frac{(n-1)(n-2)}{2 \times 1}p^2q^{n-3} + \ldots + p^{n-1}\right]$$

or

$$np\left[q^{n-1} + \binom{n-1}{1}pq^{n-2} + \binom{n-1}{2}p^2q^{n-3} + \ldots + p^{n-1}\right]$$

Inside the brackets, we have an expression which is very similar to expansion 4.14; in fact, the only difference is that $n-1$ has been substituted for n. In other words, this is another binomial expansion but with $n-1$ events being considered instead of n. This is the expansion of $(q+p)^{n-1}$, but since $q+p$ equals 1, the value of the expression is unaltered by the change in power and remains 1. So we see that the mean of any binomial distribution is always np. By similar methods it can be shown that the variance for any binomial distribution is npq or $np(1-p)$.

Understanding what is implied by the mean and variance of probability distributions, is not easy. We are now referring to probable outcomes, whereas previously, in connection with frequency distributions (see Chapters 2 and 3), we were talking in terms of many observations whose values were known. Returning to the hospital example, the simplest way of looking at the problem is to consider an indefinitely large number of hospital wards with five mothers-to-be in each of them. The mean of the probability distribution of the number of male births corresponds to the mean number of male births born in all the hospitals, i.e. the total number of male births divided by the number of hospitals. For probability distributions the arithmetic mean is most often termed the *expectation* or *expected value* of the variable r.

Table 4.4 *Male Births to Twenty Mothers*

r	$Pr(r)$	r	$Pr(r)$	r	$Pr(r)$
0	0.00000042	7	0.057	14	0.050
1	0.0000091	8	0.101	15	0.022
2	0.000094	9	0.146	16	0.0074
3	0.00061	10	0.173	17	0.0019
4	0.0028	11	0.171	18	0.00034
5	0.0098	12	0.139	19	0.000039
6	0.026	13	0.093	20	0.0000021

When a graph is drawn, the binomial distribution can be seen to possess a shape which varies according to the values of p and n. When p is at or near 0.5 (as for coin tossing or the sex of babies), and n is greater than (about) 10, then the binomial has a shape which is quite close to that of the normal curve. For example, the probabilities of the values of r (number of male births) for 20 mothers-to-be (assuming no multiple births) are shown in Table 4.4 and the shape of the distribution is indicated in

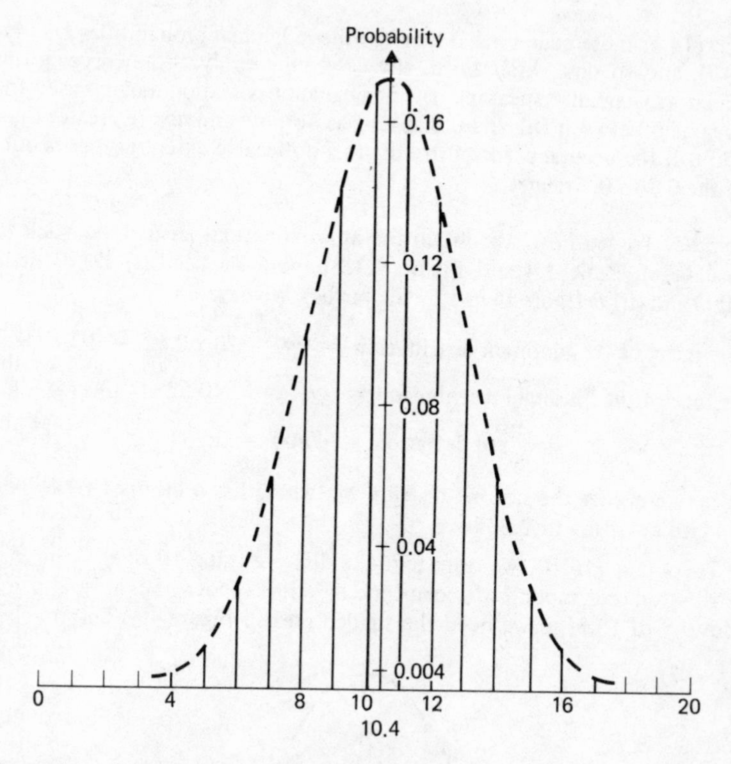

Figure 4.3 *Binomial distribution: male births.*

Figure 4.3, again taking $p = Pr(\text{male birth}) = 0.52$. Although not perfectly symmetrical, the distribution indicated by the broken line is obviously very close to the normal curve. (When dealing graphically with discrete distributions like the binomial, one must use broken lines to show the shape, for continuous lines might be taken to imply that all values were possible whereas, in fact, only a very limited number of values is obtainable, i.e. those indicated by the vertical lines.)

We noted in Chapter 3 that the normal distribution is often used to approximate other distributions when the calculations with those distributions get very involved. Now we can illustrate this in the case of the binomial distribution. In this connection it is necessary to give guidance as to when one should or should not use the normal to approximate the binomial distribution. There is little advantage in using it when n is less than 20 as the exact binomial probabilities can be obtained for such values quite easily with a calculator. What one can say is that the further p departs from 0.5, the larger n must be to compensate. One should also distinguish between problems where one needs the probability of particular values and those where one requires the probability of a set of values. For example, for the data of Table 4.4 we would obtain a better approximation for Pr(14 or more male births) than for the individual probabilities $Pr(14)$, $Pr(15)$, and so on, which are in any case quite easily obtained using the calculator. Generally speaking, the normal approximation may be used for values of p between 0.1 and 0.9 as long as both np and nq are greater than 5, though the accuracy does drop off to a noticeable extent as p goes outside the 0.3 to 0.7 range.

Example. To see how we obtain the approximation, let us refer back to the data of Table 4.4 and Figure 4.3. Suppose we need (i) Pr(10 male births) and (ii) Pr(more than 12 male births). We have:

$$\text{mean of the binomial distribution} = np = 20 \times 0.52 = 10.4$$

$$\text{variance of the binomial distribution} = npq = 20 \times 0.52 \times 0.48 = 4.992$$

$$\text{standard deviation} = \sqrt{npq} = 2.234$$

We now consider the normal distribution which has a mean of 10.4 and a standard deviation of 2.234:

(i) To obtain $Pr(10)$, we note that the discrete value 10 of the binomial distribution corresponds to continuous values under the normal curve from 9.5 to 10.5, so we need the shaded area in Figure 4.4. This involves

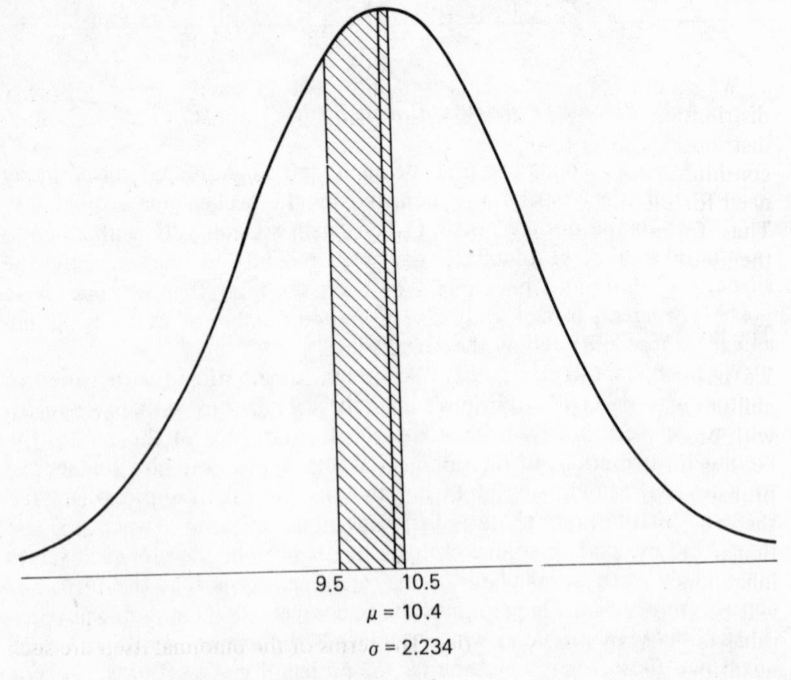

Figure 4.4 *Probability of ten male births.*

finding the z-values corresponding to 9.5 and 10.5:

$$z_1 = \frac{9.5-10.4}{2.234} = -0.403 \qquad z_2 = \frac{10.5-10.4}{2.234} = 0.045$$

From Table A in the Appendix containing normal areas, we find area below z_1 is 0.3435, area below z_2 is 0.5180. If we subtract the smaller area from the larger, we obtain the probability corresponding to the value 10:

$$Pr(\text{10 male births}) = 0.5180-0.3435 = 0.1745$$

The exact value for $\binom{20}{10}(0.52)^{10}(0.48)^{10}$ is 0.1734, so our approximation is well within 1 per cent of the true value.

(ii) More than 12 male births implies 13 or more. Again, the binomial value 13 must be taken as corresponding to normal values from 12.5 to 13.5. In this problem, we focus our attention on 12.5:

$$z = \frac{12.5-10.4}{2.234} = 0.940$$

From Table A in the Appendix, we find that this corresponds to an area of 0.8264, so

$$Pr(\text{more than 12 male births}) = 1-0.8264 = 0.1736$$

We should note at this point that, although we use the standard normal distribution to obtain approximate values for probabilities from other distributions, it is itself a probability distribution in its own right. As a continuous distribution it has a total area of 1, which corresponds to the need for the individual probabilities of a discrete distribution to sum to 1. Thus, for certain purposes, the areas under the standard normal curve may themselves be treated as probabilities.

The Poisson Distribution

We have now dealt with the problem of approximating binomial probabilities where p lies between 0.1 and 0.9, but quite often we have to deal with problems where the probabilities of events happening are very small, i.e. less than 0.1. Accident data provide good examples of such events. The probability of a person being killed on our roads in a particular year is of the order of 1:7,000. The probability of a person being injured in a particular year is of the order of 1:200. Suppose that in a town of 20,000 inhabitants we want to know the probability that more than five people will be killed in a particular year. The normal approximation is not applicable for such small probabilities. The terms of the binomial itself are such quantities as

$$\binom{20{,}000}{4}(1/7{,}000)^4(6{,}999/7{,}000)^{19{,}996}$$

which is the probability of exactly four persons being killed. Although such expressions are certainly not beyond the scope of even quite modest modern calculators (e.g. the above quantity can be shown to be 0.1595), they can still be quite time-consuming when there are a number of them. Luckily there is a distribution which approximates the binomial very well under such circumstances, called the Poisson distribution.

It can be shown that when n is large and p very small and np is represented by a, then the terms of the binomial distribution given by the expansion 4.14 are approximated by those of the series

$$\frac{1}{e^a}+\frac{a}{e^a}+\frac{a^2}{2!e^a}+\frac{a^3}{3!e^a}+\ldots$$

which itself sums to 1 and, therefore, represents a probability distribution. The approximation is a good one as long as $a<5$. The quantity e is well known in mathematics and is called the exponential constant. Its numerical value is 2.71828 correct to five decimal places. The above series of terms comprises the values of the *Poisson distribution* . The probability of obtaining a particular whole number value r for this distribution is given by

$$Pr(r) = \frac{a^r}{r!e^a} \quad \text{or alternatively} \quad \frac{a^r}{r!}e^{-a} \tag{4.15}$$

Example. In the case of the problem posed on road accidents, we have $n = 20{,}000$, $p = 1/7{,}000$, so $a = np = 20{,}000/7{,}000 = 2.857$. It follows that $e^a = e^{2.857} = 17.41$ and $e^{-a} = 1/e^a = 0.0574$. To obtain an approximate value for the binomial probability given earlier, i.e. the probability of four persons being killed, we have from expression 4.15

$$Pr(4) = \frac{2.857^4 e^{-2.857}}{4!} = 0.1595$$

which is identical to the binomial value even to the fourth decimal place.

To find the probability that more than five people are killed, we again make use of the fact that the sum of all the probabilities is 1. Hence

$$\begin{aligned} Pr(\text{more than }5) &= 1 - Pr(5\text{ or fewer}) \\ &= 1-(Pr(0)+Pr(1)+Pr(2)+Pr(3)+Pr(4)+Pr(5)) \end{aligned}$$

Putting r successively equal to 0, 1, 2, 3, 4 and 5 and working to three decimal places, we obtain from expression 4.15:

$$Pr(0) = e^{-2.857} = 0.057 \qquad Pr(1) = 2.857e^{-2.857} = 0.164$$

$$Pr(2) = \frac{2.857^2 e^{-2.857}}{2!} = 0.234 \qquad Pr(3) = \frac{2.857^3 e^{-2.857}}{3!} = 0.223$$

$$Pr(4) = \frac{2.857^4 e^{-2.857}}{4!} = 0.159 \qquad Pr(5) = \frac{2.857^5 e^{-2.857}}{5!} = 0.091$$

The sum of these probabilities is 0.928. Thus, the probability of more than 5 deaths is $1 - 0.928 = 0.072$. This result is exactly the same as would be obtained from the binomial distribution, to the third decimal place.

Because the relationship between successive Poisson probabilities is very simple, namely,

$$Pr(r+1) = \frac{a}{r+1} Pr(r)$$

their determination from $Pr(0)$ is particularly easy, i.e.

$$Pr(1) = \frac{a}{1} Pr(0) \quad Pr(2) = \frac{a}{2} Pr(1), \text{ and so on}$$

These relationships can be used to make an almost immediate calculation of Poisson probabilities, using a modern calculator.

The Poisson distribution, while being very useful as an approximation to the binomial distribution, is important in its own right. It describes very well the observed distribution of variables, such as the number of goals scored in football matches, the number of people contracting certain rare diseases, the traffic density on roads, and the frequency of thunderstorms; in fact, almost any event which is unpredictable in its nature and where the probability of occurrence is very small in a particular time interval. It is also one of the very few statistical distributions for which the mean and the variance are equal. This fact serves to distinguish it in practice from the binomial.

Glossary

Probability
Conditional probability
Independent events
Dependent events
Mutually exclusive events
Sampling with replacement
Sampling without replacement
Probability distribution
Expectation (or expected value)
Binomial distribution
Poisson distribution

Exercises

1 The probability that a man aged 45 will be alive in 20 years time was calculated on certain assumptions to be 0.7 and that of his wife – also aged 45 – to be 0.8. Assuming the occurrence of each death is independent, obtain the following probabilities: that (a) both and (b) neither

will be alive in 20 years time; that (c) only the man and (d) only his wife will be alive then; that (e) at least one person will be alive.

2 For a family with four children, what is the probability that there are (a) no boys, (b) two boys and two girls, (c) at least one boy, (d) at most two boys? Assume that boys and girls are equally likely and that no control has been exercised over the sex composition of the family.

3 In the adult population of a region 28 per cent of persons are single (never having been married), 55 per cent are married, 10 per cent are divorced and a further 7 per cent widowed. What are the probabilities among ten adults who have assembled independently and by chance that there are (a) at least two divorced or widowed persons, (b) exactly three single adults, (c) at most five married persons, (d) at least nine persons who have had marital experience?

4 If there is a 1:1,000 chance of a birth resulting in a handicapped child, what is the probability that 4,700 births give rise to the following numbers of handicapped children (assuming no multiple births and that all births are equally likely to result in handicap): (a) none, (b) at least two, (c) no more than six? Which number has the highest probability?

5 It is estimated that in Wales there is a probability of 0.04 of a motor cyclist incurring injury in (one or more) road accidents during a year. Supposing individuals are equally likely to be injured in this way, find for a group of 200 motor cyclists the probability that the following numbers will be injured in a particular year: (a) fewer than four, (b) five, (c) more than ten. Determine binomial probabilities to obtain your answers and also calculate the Poisson and normal approximations and compare them (scientific calculator definitely required).

Chapter 5

Sampling and Estimation

Statistics is most often concerned with the analysis of sample data, because the direct measurement of entire populations is usually impracticable or actually impossible. We shall now describe how samples are taken from populations – a very important part of statistics, about which many books have been written (see Moser and Kalton, 1971, chapters 4–7; Cochran, 1953), though we shall only be concerned here with the most basic methods.

As explained in Chapter 1, a central purpose of recording and analysing sample data is to estimate population parameters such as the mean μ and the variance σ^2, using quantities calculated from the sample. These latter quantities are known as sample statistics. Because samples necessarily contain only some of the values from the population, estimates based on them must be subject to error. In other words, we are unable to guarantee that the estimate we obtain from the sample data will be equal to, or even very close to, the population parameter we are estimating. In this connection the magnitude of the difference between the sample statistic and the parameter it is estimating is of vital interest to us. As the parameter's value will be unknown, the difference between statistic and corresponding parameter cannot be exactly determined, and we have to rely on probabilistic methods. These methods can only be used if the technique of sampling adopted incorporates, to a greater or lesser extent, an element of random selection.

Sampling Methods

A random method of sample selection is such that each member of the population has a known, non-zero, probability of being included in the sample. We have already described, without actually naming, the most basic method of random sampling when we talked, in Chapter 4, about selecting a club committee by drawing a sample from some numbered slips shuffled in a container. This exemplifies a scheme termed *simple random sampling*, under which each possible sample of the chosen size n drawn from a population of size N has the same probability of being selected (where n is less than N). Given that sampling is carried out without replacement there are $\binom{N}{n}$ essentially different samples, any one of which has a probability of $1\Big/\binom{N}{n}$ of being selected. It also follows that each member of the population has exactly the same probability of being included in the sample. The results which we shall develop will be based on the assumption that such a scheme is being used and whenever in later chapters we refer to random sampling, it is to be taken to be simple random sampling (or, alternatively, a scheme considered equivalent to it, e.g. quasi-random sampling, discussed below).

In practice, especially where large populations are involved, simple random sampling may not be very easy to achieve. It requires that the entire population be available for sampling purposes, in the form of a list or map, etc. The latter is what is known as a *sampling frame.* Two of the most useful sampling frames in the social sciences are the electoral roll and rating lists, which enable us to obtain samples of the electorate and dwelling-places, respectively. However even these lists present a problem, for they are never completely up to date. With small populations such as first-year students at a university or employees of a firm, simple random sampling is generally practicable, for suitable lists are usually available within the organisation itself, or alternatively, the investigator may well be able to compile a list himself.

Having satisfied oneself that the sampling frame is suitable, two basic methods of selection are available which ensure simple random sampling. The first, the *lottery method*, has already been noted and essentially involves selecting the required number of population members by drawing from numbered slips shuffled in a box (or, equivalently, one might use dice instead). The second involves the use of *random numbers.* A table of random numbers will be found in the Appendix of this book (Table H). It will be observed that the random numbers are in five-digit form, in blocks of five, and the blocks are organised into ten rows and ten columns on each page. To use the table in connection with sample selection, the members of the population need to be numbered from 1 to N and n numbers are then selected from the table in the following way. Suppose that n is 50, while N is 600. Since N has three digits, we use three adjacent columns of the table, proceeding to the neighbouring three columns after reaching the bottom of the page. If we start with the first three columns of Table H, then the first number which we select into the sample is the first number between 001 and 600 which appears, i.e. 494. There then follow numbers 294, 252, 024. The next number would be 694, but we reject this because it exceeds N. We continue in this way down the page only taking numbers less than or equal to N. We also reject any numbers that have already been selected, since we are sampling without replacement. The simple random sample has been selected when we have chosen 50 three-digit numbers corresponding to 50 population members.

Where simple random sampling is inconvenient or inappropriate, other methods of selection embodying random features may be chosen. One which is particularly useful is *stratified random sampling*, which makes use of knowledge about the population, frequently to make the sample more representative and generally to increase the probable accuracy of estimates derived from it. To illustrate the use of stratified random sampling, consider the following situation. We wish to investigate the amount of time which married people are able to devote to leisure activities. Previous research suggests that men and women differ considerably in the time they have available for such activities. Simple random samples from an adult population would contain varying numbers of men and women. What we seek to do is to represent each sex in the sample in the same proportion as in the population. We achieve this by splitting the population into men and women and taking a simple random sample of each sex. The splitting is usually easily effected, since lists of people generally enable males and females to be distinguished either by title or by first name. Finally, the two simple random samples are combined and we have our single stratified random sample. In general there may not simply be two, but any number of strata. When the same proportion of each stratum is taken into the sample, we have a *proportionate stratified sample*, otherwise it is *disproportionate*. Whenever there is a heterogeneous population which can, without too much effort, be split into a number of relatively homogeneous groups or strata, then stratified random sampling should be used.

A further sampling method which is eminently practicable, although not strictly random, is *quasi-random sampling* or *systematic sampling from lists*. This is a scheme used widely where the sampling frame is in an order unrelated to the subject of the inquiry (e.g. alphabetical order). If we have a population with N members and we seek to obtain a 1-in-20 sample or, as we say, the *sampling fraction* is 1:20 or 5 per cent, then we select one member from the first 20 by some random method (e.g. random numbers) and then take every twentieth member from the list until we reach the end. Thus, if we have a list of 987 people and we want a 5 per cent sample, we select perhaps the 14th, then the 34th, 54th, 74th persons, and so on, to the 954th and 974th persons, when our sample is complete. It must be noted that in this instance had the first number chosen happened to be 7 or less, there would have been one additional person in the sample. This slight problem will always arise when the population is not an exact multiple of the sampling fraction but unless the sample size is very small it is of little consequence. What we obtain by this method is a sample selected with a random element. However, we do *not* have a simple random sample, for although the various samples which we might have drawn do have the same probability of being selected, their number is severely restricted compared with that of a simple random sampling scheme. In the latter there are $\binom{N}{n}$ possible samples, whereas the quasi-random scheme has approximately N/n, or to be precise the reciprocal of the sampling fraction which was 20 in the example described. Each member of a quasi-random sample is dependent on the previous selection, whereas every member of a simple random sample is selected independently. Does this mean that the results we shall develop cannot be applied to quasi-random samples? Not really. As long as care is taken in choosing the list so as to ensure that there is no trend or cyclical movement related to the subject of study, one can treat such samples as if they were simple random samples. Indeed, quasi-random sampling is used on a very large scale in practice and the saving in effort as compared with using simple random sampling is great.

At this point in our consideration of sampling, we need to define formally certain important terms. Suppose that, using repeated simple random sampling, we take an indefinitely large number of samples of size n from a population of N members (replacing each sample before selecting another one). Each sample will provide an estimate of a population parameter, e.g. each sample mean will be an estimate of the population mean. Some of the sample estimates will be close to the parameter value, others will not. However, if we take the mean value of all the sample estimates (necessarily a theoretical concept), this mean value is known as the *expected value* of the sample statistic. In this context the sample statistic is known as an *estimator*. An estimator is said to be *unbiased* if its expected value equals the parameter being estimated by it. Otherwise it is *biased*. It can be shown that the sample mean is an unbiased estimator of the population mean, whereas the sample variance is a biased estimator of the population variance (both of these estimators are considered below). The *bias* is the difference between the expected value of the estimator and the population parameter. On the other hand, the *accuracy* of an individual estimate of a parameter is defined as the actual difference between that estimate and the parameter. Since in practice the parameter remains unknown, we are unable to determine the accuracy exactly and we have to talk in terms of the *probable accuracy*, i.e. the probability that the difference between the value of the estimator and the parameter is less than a given magnitude. In order to keep the latter magnitude small, it is desirable that an estimator be unbiased and also that the values which it may take tend to cluster closely round the parameter. When this latter is the case, we say that the *precision* of the estimate is high. We can further clarify the notion of precision only after we have described sampling distributions.

Sampling Distributions

In the paragraph above, we imagined taking an indefinitely large number of simple random samples of size n from a population with N members and then taking the mean of all the sample statistics derived therefrom. If we now go a step further and imagine the building up of a grouped frequency distribution for these sample statistics (in particular the sample means), the resulting distribution will be the *theoretical sampling distribution* of the statistic. The sampling distribution of the mean is of special significance to us in the work which follows. We need to know what its characteristics are, i.e. its shape, mean and variance. The distribution is specified as follows:

(1) If samples of any specified size n are drawn from a population which is *normally* distributed with mean μ and variance σ^2, then the sampling distribution of the mean is *exactly normal* with mean μ and variance σ^2/n.

(2) For samples of size n drawn from *any (large) intervally scaled population* with mean μ and variance σ^2, then the *normal distribution* with mean μ and variance σ^2/n will be a good *approximation* to the sampling distribution of the mean, provided n is *sufficiently large*. This second result is essentially a simplified statement of the very important *central limit theorem*, the precise form of which is beyond the scope of this book.

The implications of the above results are very important in regard to a great deal of the work in this chapter on estimation and in subsequent chapters on significance testing. From a practical point of view the second result is particularly vital, because it permits the use of normal curve methods in problems involving means even when the initial population has a distribution that differs considerably from normality. Just how large n must be in order that the approximation be adequate, depends on how asymmetrical the population is. For symmetrical or near-symmetrical distributions samples of size 10 to 15 may be large enough, whereas for highly asymmetrical distributions samples in the hundreds may be required. However, sampling experiments have revealed that for most populations likely to be encountered the approximation is good for $n > 50$. In using the result to estimate μ, we focus particularly on the standard deviation of the sampling distribution of the mean, i.e. $\sigma/\sqrt{n}$, which is more commonly known as the *standard error of the (sample) mean*. In the notation introduced earlier in connection with the normal curve (p. 37) the distribution statement made verbally in (1) above can be written symbolically, thus:

$$\bar{x} \sim N\left[\mu, \frac{\sigma^2}{n}\right] \tag{5.1}$$

For (2) above, we would use the symbol $\stackrel{\cdot}{\sim}$ to indicate an approximate distribution:

$$\bar{x} \stackrel{\cdot}{\sim} N\left[\mu, \frac{\sigma^2}{n}\right] \tag{5.2}$$

These results apply for infinite or large populations where the withdrawal of a sample has no appreciable effect upon the composition of the population. Where this is not the case and in particular where the sampling fraction n/N is greater than 5 per cent, the variance term in expression 5.2 must be adjusted using a *finite population correction* (f.p.c.). To be precise

Table 5.1 *A Theoretical Sampling Distribution*

$\bar{x}$	f	$\bar{x}$	f
2	1	9	1
$2\frac{1}{3}$	1	$8\frac{2}{3}$	1
$2\frac{2}{3}$	2	$8\frac{1}{3}$	2
3	3	8	3
$3\frac{1}{3}$	4	$7\frac{2}{3}$	4
$3\frac{2}{3}$	5	$7\frac{1}{3}$	5
4	7	7	7
$4\frac{1}{3}$	8	$6\frac{2}{3}$	8
$4\frac{2}{3}$	9	$6\frac{1}{3}$	9
5	10	6	10
$5\frac{1}{3}$	10	$5\frac{2}{3}$	10

the variance is reduced by multiplication by the factor $(N-n)/(N-1)$. Thus, in these circumstances expression 5.2 becomes*

$$\bar{x} \mathrel{\dot{\sim}} N\left[\mu, \frac{\sigma^2}{n}\left(\frac{N-n}{N-1}\right)\right] \tag{5.3}$$

The following example will serve to highlight these results.

Example. Suppose that our population consists of the numbers from 1 to 10 inclusive. We have mean $\mu = 5.5$, variance $\sigma^2 = 8.25$, size $N = 10$. Let us imagine taking an indefinitely large number of random samples of size three from this population and determining the sample means. Now the number of essentially different samples is not unlimited, in fact it is $\binom{10}{3} = 120$. Since we are using random sampling, in the long term these 120 samples and their associated means will occur equally often. So focusing on the essentially different samples, the frequency distribution for the means will be as shown in Table 5.1, which has been arranged so as to emphasise the symmetrical form.

The histogram shown in Figure 5.1 is very close to the curve of the normal distribution having the same mean and variance. The mean of the sampling distribution is equal to the population mean 5.5, as expected, but the variance of 2.139 differs quite substantially from the value of $\sigma^2/n = 8.25/3 = 2.75$. Why should this be? It is because we have taken a

*When the population is large (e.g. $N > 100$), it is in some respects simpler and almost always good enough to write instead

$$\bar{x} \mathrel{\dot{\sim}} N\left[\mu, \frac{\sigma^2}{n}(1-f)\right]$$

where $f = n/N$ is the sampling fraction.

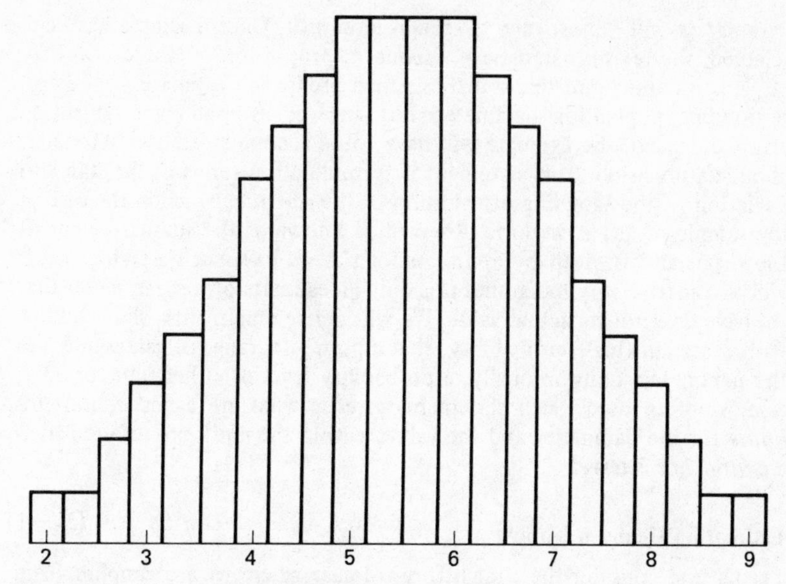

Figure 5.1 *A sampling distribution.*

substantial proportion of the population into each sample, and it is a good illustration of the effect of the f.p.c. factor, which in this case is $(10-3)/(10-1) = 7/9$. We note that

$$\frac{\sigma^2}{n}\left(\frac{N-n}{N-1}\right) = 2.75 \times \frac{7}{9} = 2.139$$

The f.p.c. does not enter into our calculations when we are dealing – as we usually are – with moderate samples from very large populations, but where the sample is a substantial proportion of the population, as in our example, then the f.p.c. has to be brought into the reckoning. In the remainder of this book, we will note the f.p.c. where it is appropriate to do so, i.e. where the sampling fraction exceeds 5 per cent. Otherwise we shall not refer to it.

What is the purpose of considering sampling distributions? Why are they so important in statistics? The answer to these questions is that a consideration of sampling distributions enables us to determine (a) whether a particular statistic used as an estimator is biased and if so the magnitude of the bias, and (b) to what extent the values of the statistic tend to cluster closely around the expected value, i.e. the magnitude of the precision of the estimator.

Estimation

We have asserted that in taking samples our purpose is frequently to obtain estimates of unknown population parameters. We stress again that these parameters will almost always remain unknown. Once a sample has been selected, the relevant statistic – the mean, proportion or whatever it may be – is obtained and this statistic is then used as an estimate of the corresponding population parameter. But since all estimates are subject to error, our particular estimate will have to be accompanied by a statement about its precision. This statement is formulated in terms of the standard deviation of the sampling distribution of the statistic, which in the case of the sample mean, as we have observed, is known as the standard error of the mean and is given by $\sigma/\sqrt{n}$. As long as we have such a statement, it will not matter very much that it is only an estimate of the parameter that we have and not its actual value. We will derive statements which tell us, with a certain (high) probability, that a particular range of values includes the parameter. Conventionally, a probability level of either 0.95, or 0.99, is most often used. Such statements give us what are called *confidence limits* for the parameter and the values within the limits are referred to as a *confidence interval*.

Estimating Means, σ Known

Let us first consider the straightforward case where we are sampling from a population (not necessarily normal) with mean μ and variance σ^2, where σ is known. We seek to estimate μ, using $\bar{x}$. Now if the sample size is sufficiently large, we are assured that result 5.2 holds true, i.e. that the sampling distribution of the mean is approximately normal with mean μ and variance σ^2/n. It follows that

$$\frac{\bar{x}-\mu}{\sigma/\sqrt{n}} \tag{5.4}$$

is approximately distributed as z, the standardised normal variable.

Referring to Table A in the Appendix, we can determine that 0.95 or 95 per cent of the area under the standard normal curve lies between z-values of -1.96 and $+1.96$. Applying this to the distribution of sample means, we are able to say that with repeated sampling 95 per cent of the sample means will have values within $1.96\sigma/\sqrt{n}$ (or 1.96 standard errors) of the population mean and 5 per cent will not. If we now construct intervals based on these sample mean values such that the upper limits are $1.96\sigma/\sqrt{n}$ above and the lower limits are $1.96\sigma/\sqrt{n}$ below the sample means, then in 95 per cent of cases μ will be contained in the interval and in 5 per cent it will not. Referring to Figure 5.2, we can say that for any sample mean $\bar{x}$ lying between L and U (e.g. $\bar{x}_1$) μ will be contained in the interval constructed about $\bar{x}$, otherwise it will not (e.g. $\bar{x}_2$). It is these intervals which are referred to as confidence intervals and their end-points are confidence limits. In this case, we say that we have 95 per cent confidence limits. A particular value of the sample mean $\bar{x}$ is our estimate of μ – called a *point estimate* – but it is subject to error. What we need to specify is the magnitude of that error. The above argument assures us

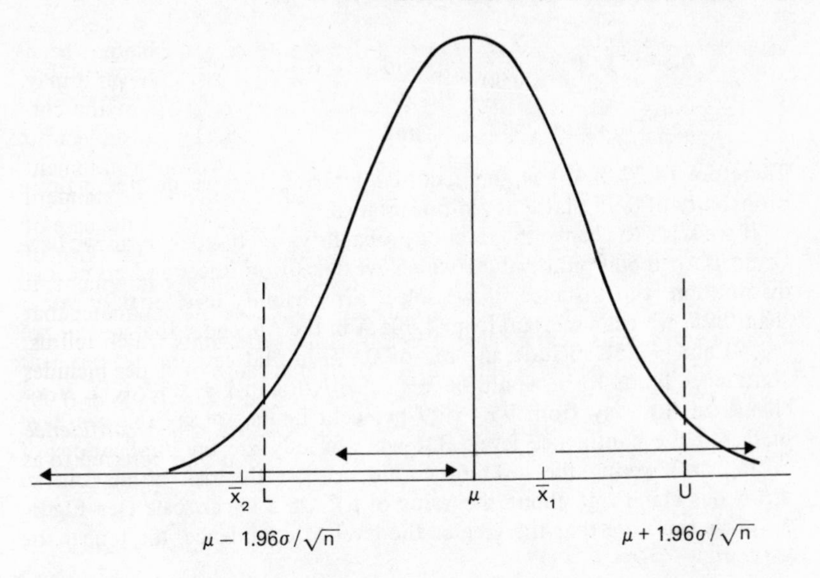

Figure 5.2 *Confidence intervals.*

when taking an indefinitely large number of random samples from a given population that in 95 per cent of cases the population mean will be contained in an interval defined as being from $1.96\sigma/\sqrt{n}$ below to $1.96\sigma/\sqrt{n}$ above the sample mean.

In practice, of course, we will generally be dealing with a *single sample* and from it we will want to make a meaningful statement about the value of the population mean. What we may be tempted to say, having obtained the particular value of the sample mean, is that there is a probability of 0.95 that μ will lie within $1.96\sigma/\sqrt{n}$ of it. But can we make such a statement? Some statistics books imply that we can. However, although μ is unknown, and is likely to remain so, it does have a fixed value. Therefore, once we have constructed a confidence interval about a particular sample mean, either μ will be in that interval or it will not be. So the kind of probability statement which we have proposed above is not possible. How then can we express our ideas about the value of μ? We can put them this way. If we state that μ is contained in the interval $\bar{x} - 1.96\sigma/\sqrt{n}$ to $\bar{x} + 1.96\sigma/\sqrt{n}$, then there is a probability of 0.95 that the *statement* is correct (and a probability of 0.05 that it is incorrect). Alternatively, if we use a value of $\bar{x}$ as a point estimate of μ, there is a probability of 0.95 that the error we may be committing will not be greater than $1.96(\sigma/\sqrt{n})$.

Example. If $\bar{x} = 6.34$, $\sigma = 10.56$ and $n = 100$, specify a point estimate of μ and determine a 95 per cent confidence interval. The value of $\bar{x}$, 6.34, is the required point estimate; 95 per cent confidence limits are

$$6.34 - 1.96 \times \frac{10.56}{\sqrt{100}} \quad \text{and} \quad 6.34 + 1.96 \times \frac{10.56}{\sqrt{100}}$$

$$\text{i.e.} \quad 4.27 \quad \text{and} \quad 8.41$$

Therefore (4.27, 8.41) is the required interval, but we do *not* claim a probability of 0.95 that μ lies in this interval.

If we wish to change the level of probability, or confidence, all we have to do is to modify the values which we take from the standard normal distribution. For instance, if we select a probability level of 0.99 rather than 0.95, we can ascertain from Table A in the Appendix that z-values of -2.58 and $+2.58$ include an area of 0.99. In that case our 99 per cent confidence limits for μ would be $\bar{x} - 2.58\sigma/\sqrt{n}$ and $\bar{x} + 2.58\sigma/\sqrt{n}$. If we change in this way from 95 to 99 per cent limits, we say that we have increased the confidence level. But what exactly are the implications of raising or lowering the level of confidence (or the risk of making an erroneous statement about the value of μ)? On a linear scale (see Figure 5.3), we can note that the greater the level of confidence, the longer the

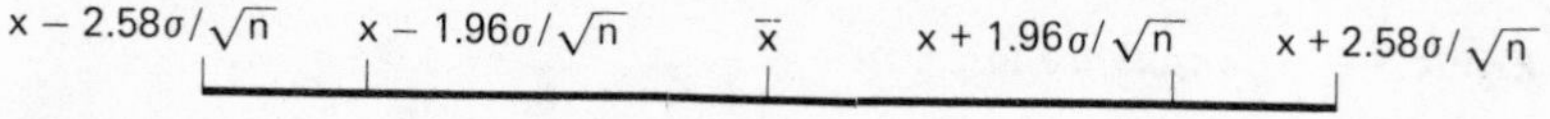

$x - 2.58\sigma/\sqrt{n}$ $\quad$ $x - 1.96\sigma/\sqrt{n}$ $\quad$ $\bar{x}$ $\quad$ $x + 1.96\sigma/\sqrt{n}$ $\quad$ $x + 2.58\sigma/\sqrt{n}$

Figure 5.3 *Levels of confidence.*

interval has to be, for the same size of sample. For instance, in our earlier example a 99 per cent confidence interval for μ (as the reader may care to confirm) is (3.62, 9.06) as opposed to the 95 per cent interval of (4.27, 8.41). Therefore, there has to be a compromise between confidence and the length of the interval. The shorter we require the confidence interval to be, the lower the probability or confidence level. There is only one way we can have both tighter confidence limits, and a higher level of confidence with a given population distribution, and that is by increasing the value of n. The larger the sample, the narrower the limits and/or the higher the level of confidence. But one must note well that the standard error, which is the fundamental indicator of precision, is only reduced by the square root of the factor by which the sample size is increased. So in order, at a given confidence level, to halve the length of the confidence interval, we have to quadruple the sample size. Although the latter will probably not quadruple the total time and cost of data collection, it must still increase them quite substantially. Once again, there is a need for compromise. In connection with an investigation it is necessary to indicate the minimum degree of precision which is required and then allocate sufficient resources to data collection in order to achieve it.

Estimating Means, σ Unknown

So far we have assumed that the population variance σ^2 is known. In practice, if we do not know the population mean it is only very rarely that we will know the variance. So how do we proceed? We have $\bar{x}$ as an estimator of μ, but we seem unable to determine a confidence interval. In this event, we have to obtain an estimate of σ^2 from the sample data. This brings in the additional problem as to which sample statistic we should use as an estimator of σ^2. Our immediate reaction might be to use the sample variance S^2 defined in expression 3.4 (p. 29). However, it can be shown that this is not the best procedure. When we talked about the sampling distribution of the mean, we said that if we took all possible samples of size n from a population with mean μ and determined the overall mean of all the means of those samples, the resulting quantity would be μ. In more technical language the expected value of the sample mean $\bar{x}$ is the population mean μ, and $\bar{x}$ is said to be an unbiased estimator. If we repeat this process for sample variances, we find that the mean of the sample variances is not σ^2, i.e. the expected value of S^2 is not σ^2 but a value somewhat below it. So S^2 is a biased estimator, and as such is not wholly satisfactory in estimating σ^2. An unbiased estimator is, in fact, obtained by dividing the sum of squares of deviations by $n-1$ rather than n. The resulting quantity

$$s^2 = \frac{\Sigma(x-\bar{x})^2}{n-1} \tag{5.5}$$

is unbiased in that its expected value is σ^2. Henceforward we shall refer to S as the *unadjusted standard deviation* of the sample, while s is the *adjusted standard deviation* and in work on estimation and hypothesis testing we almost invariable use the latter quantity.

The number $n-1$ is of special significance in this context. It is referred to as the number of *degrees of freedom* associated with the sum of squares of deviations and is given the symbol ν (Greek nu). Some explanation of the terminology may be helpful. Suppose we have a sample with (say) six values, namely, x_1; x_2; x_3; x_4; x_5; x_6. These x-values can each take on any numerical value within the range offered by the population. We can say that each has a degree of freedom. Let us now consider the deviations of the x-values from their mean $\bar{x}$. We know that the sum of the deviations is always zero (see p. 28). So, if we omit the last deviation $(x_6-\bar{x})$ and sum the other five, it is evident that the deviation $(x_6-\bar{x})$ must be the difference between zero and the total of the other five. In other words, once $n-1$ deviations have been determined, each having a degree of freedom, the nth is determined by the difference between 0 and the previous $n-1$. Hence the deviation 'has no freedom' to take on any other value, and we say that one degree of freedom has been lost (compared with the initial n). Whenever we are dealing with measures based on deviations from the mean, we have only $n-1$ degrees of freedom. We use the concept of degrees of freedom at various points in this and subsequent chapters.

However, let us now address ourselves directly to the problem of estimating the mean of a population when σ is unknown. We have established in this connection that s^2 is the appropriate estimator of σ^2. Using this result, it turns out that for *normal or approximately normal populations* we are able to specify confidence intervals for μ. We have seen (result 1, p. 73) that when the population is normally distributed, the statistic $\bar{x}$ is (exactly) normally distributed with mean μ and variance σ^2/n. It follows that

$$z = \frac{\bar{x} - \mu}{\sigma/\sqrt{n}}$$

conforms to the standard normal distribution. In order to tackle the problem of obtaining a confidence interval when σ is unknown, let us consider instead the apparently similar but importantly different variable quantity

$$t = \frac{\bar{x} - \mu}{s/\sqrt{n}}$$

This is not normally distributed but is distributed as *Student's t-distribution* (named after W. S. Gossett, who wrote under the pseudonym 'Student'). The t-distribution is similar to the standard normal in certain respects, e.g. it is symmetrical about zero, but it is flatter and has proportionately more of its area in its tails. The reason for this is that s varies from sample to sample just as $\bar{x}$ does. This implies that the quantity t will be more variable than z.

The t-distribution is in fact not a single distribution, but a family of distributions. The particular distribution of t depends on $n - 1$, the number of degrees of freedom (ν). For small values of ν the t-distribution is somewhat flatter than the standard normal (Figure 5.4), but for large values of ν (say over 50) the two distributions are very close to one another. As ν becomes indefinitely large ($\nu = \infty$), the two distributions become identical. Values of the t-distribution are given in Table B in the Appendix at the end of the book. Because the values of t vary according to the number of degrees of freedom, this table is organised differently from the table of normal areas (Table A, Appendix). In the left-most column values of ν are listed. In the following columns values of t are given such that 0.05, 0.025, 0.01, 0.005, 0.001 and 0.0005 of the area lies above them (the total area under the curve being 1). Given the distribution's symmetry, corresponding values in the lower (left) half of the distribution can be obtained by changing the sign of the t-value. Thus with 9 degrees of freedom, 0.025 (or 2.5 per cent) of the area comes above the t-value of 2.262 and also below a t-value of -2.262. So these two values include 0.95 (or 95 per cent) of the total area. Similarly with 19 degrees of freedom, t-values of 2.861 and -2.861 include 0.99 (99 per cent) of the area.

We can now obtain confidence limits for μ in much the same way as we did when σ was known, the difference being that we now use the appropriate t-values instead of z. For instance, from the above figures we can state that if $n = 10$ and hence $\nu = 9$, $\bar{x} - 2.262s/\sqrt{10}$ and $\bar{x} + 2.262s/\sqrt{10}$ are 95 per cent confidence limits for μ. Let us illustrate the use of the t-distribution in a small sample problem.

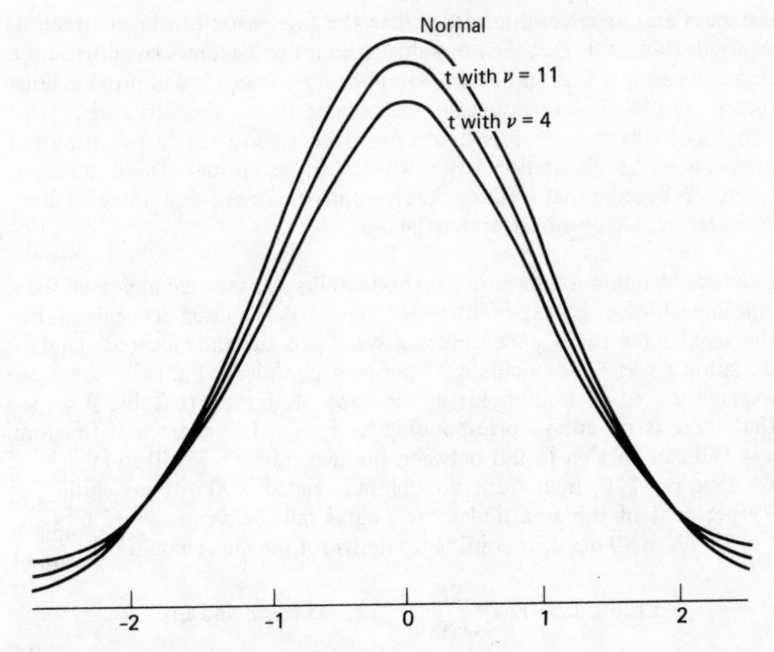

Figure 5.4 *Standard normal and* t *distributions.*

Example. An IQ test is given to a random sample of students entering a comprehensive school. Assuming that IQ is a normally distributed variable, obtain 95 per cent confidence limits for the mean IQ of all students entering the school. The results are:

95 108 131 101 107 92 76 111 83 88 105 124 79 96 102

We note that $n = 15$ and find that $\bar{x} = 99.87$ and $s = 15.482$. Referring to Table B in the Appendix with 14 degrees of freedom, we observe that an area of 0.025 falls beyond a t-value of 2.145. Therefore, our 95 per cent confidence limits for the mean IQ are

$$99.87 \pm 2.145 \times \frac{15.482}{\sqrt{15}}, \quad \text{i.e.} \quad 91.3 \text{ and } 108.4$$

Our example assumes a population distribution which is normal. As noted above, for the t-distribution to be applicable the population must be normal or approximately normal. This statement, however, needs elaboration. Undoubtedly with small samples the requirement of near-normality is fairly strict and the number of instances in the social sciences where we have good reason to believe it is met, is rather limited. Perhaps the most common cases cited in the literature are those intelligence and attainment test scores (illustrated by our example) which are so constructed that they do yield a normal distribution for a defined population. However, these instances are the exception rather than the rule. What can be confidently asserted, though, is that the normality assumption becomes less critical the larger the sample size. In practice samples larger than 30, will provide satisfactory results despite noticeable departures from normality. With large samples of 100 or more, it is almost invariably safe to use the t-distribution even when the population is known not to be normal. These are cases where ν is so large that t-values closely approximate corresponding z-values, so where necessary the latter may be used.

Example. A random sample of 200 households was selected in a large town and in each case the expenditure per month on food was determined. For the sample the mean expenditure $\bar{x}$ was £136 and the adjusted standard deviation s was £28. Calculate 99 per cent confidence limits for the mean expenditure of all households in the town. Referring to Table B we see that there is no entry corresponding to $n-1=199$ degrees of freedom; $\nu=199$ can be seen to fall between the two entries $\nu=150$ and $\nu=\infty$. If we take $\nu=150$, then from the column headed 0.005 we conclude that 99 per cent of the area under the t-curve falls between $t=-2.609$ and $t=2.609$, so 99 per cent confidence limits for the mean would be

$$£136 \pm £2.609 \times \frac{28}{\sqrt{200}}, \quad \text{i.e.} \quad £130.8 \text{ and } £141.2$$

If instead we take $\nu=\infty$, then from the column headed 0.005 we conclude that 99 per cent of the area under the t-curve falls between $t=-2.576$ and $t=2.576$. These values are exactly what we would expect, for they correspond precisely to the corresponding z-values. In this case 99 per cent confidence limits for the mean would be

$$£136 \pm £2.576 \times \frac{28}{\sqrt{200}}, \quad \text{i.e.} \quad £130.9 \text{ and } £141.1$$

It can be readily judged that in this large sample problem it is immaterial whether we use t- or z-values. Since the specification of $\bar{x}$ involves rounding, we should report the 99 per cent confidence limits as £131 and £141.

Estimating a Proportion

Turning from the problem of estimating means to that of estimating proportions, we need to refer back to results presented in Chapter 4 on the binomial distribution. We saw there that, if a given event has a probability p of occurring in a single trial, then the probability of r occurrences of the event in n trials was given by the binomial probabilities expressed in definition 4.13. We determined also that the mean (or expected value) of the binomial probability distribution was np and the variance was $np(1-p)$ or npq. We also saw that when n was fairly large and p not too small, r was approximately normally distributed with the same mean and variance.

Now suppose that one is taking a random sample of size n from a large population which includes a proportion p of cases of a particular kind. Then if the sample is small relative to the population, each sample member has an approximate probability p of being of the specified kind. It follows that the earlier result may be applied and that the sampling distribution of r, the number of cases of the specified type in the sample, will be approximately normal with mean np and variance npq.

In using this result, one must notice, however, that in most social surveys it is not r that is of major interest to us, but r/n, the proportion of cases that occurs. For example, in a sample survey concerned with social conditions, we are generally not so much interested in knowing that 183 out of 275 houses had inside toilets as in the statistic that 0.665 or 66.5 per cent of the houses surveyed had inside toilets. The information imparted is virtually the same, but it is the proportion or percentage which is most useful for comparative work. What we now need to specify is the sampling distribution of r/n, the proportion observed in the sample.

From the earlier results, we can quickly infer that the mean or expected value of the sample proportion will be the proportion in the population, so r/n is our point estimate of p. In addition, the variance of the distribution of sample proportions is pq/n.* The square root of this quantity is the standard deviation or what is known as the *standard error of the proportion*. The distribution statement becomes

$$\frac{r}{n} \mathrel{\dot{\sim}} N\left(p, \frac{pq}{n}\right) \tag{5.6}$$

The approximation is a good one as long as $np > 5$ when $p \leqslant \frac{1}{2}$, and $nq > 5$ when $p > \frac{1}{2}$. Where the sampling fraction exceeds 5 per cent, then it becomes necessary to apply a finite population correction and the above statement is modified thus:

$$\frac{r}{n} \mathrel{\dot{\sim}} N\left[p, \frac{pq}{n}\left(\frac{N-n}{N-1}\right)\right] \tag{5.7}$$

Now that we know the sampling distribution of r/n, we can obtain confidence limits for p in exactly the same way as we did for μ. Where the f.p.c. is not needed, 95 per cent limits are given by

$$\frac{r}{n} - 1.96\sqrt{\frac{pq}{n}} \quad \text{and} \quad \frac{r}{n} + 1.96\sqrt{\frac{pq}{n}}$$

Note, however, that the parameter we are estimating, i.e. p, appears in the standard error. This forces us to use in our calculation the sample estimate

*At this point, we are using the general result that if the values of a probability distribution are all multiplied by a constant a, then the variance of the resultant distribution is a^2 times that of the original one, i.e. var.$(ar) = a^2$ var.(r). We know that for the binomial distribution var.$(r) = npq$. It follows that var.$(r/n) = (1/n^2)$var.$(r) = pq/n$.

r/n instead of p. Obviously, this must have some effect on the accuracy of the approximation but generally the effect is slight. The 95 per cent limits for p (unmodified by the f.p.c.) become

$$\frac{r}{n} - 1.96\sqrt{\frac{\frac{r}{n}\left[1 - \frac{r}{n}\right]}{n}} \quad \text{and} \quad \frac{r}{n} + 1.96\sqrt{\frac{\frac{r}{n}\left[1 - \frac{r}{n}\right]}{n}}$$

Example. Each member of a random sample of 900 electors drawn from a constituency was asked how he or she would vote, if there were a General Election the following day. The support for each political party was as follows: Labour 34.1 per cent; Conservative 36.9 per cent; Liberal 11.8 per cent; Others 7.3 per cent; Don't knows 9.9 per cent. Obtain 95 and 99 per cent confidence limits for the percentage support in the constituency electorate for each of the three major parties.

We have $n = 900$ and expressed as proportions the support for each major party was Labour, 0.341; Conservative, 0.369; Liberal, 0.118. Considering Labour support, we have $r/n = 0.341$, $(1 - r/n) = 0.659$ and the 95 per cent confidence limits are

$$0.341 - 1.96\sqrt{\frac{0.341 \times 0.659}{900}} \quad \text{and} \quad 0.341 + 1.96\sqrt{\frac{0.341 \times 0.659}{900}}$$

i.e. 0.310 and 0.372

The 99 per cent limits are

$$0.341 - 2.58\sqrt{\frac{0.341 \times 0.659}{900}} \quad \text{and} \quad 0.341 + 2.58\sqrt{\frac{0.341 \times 0.659}{900}}$$

i.e. 0.300 and 0.382

Reverting to percentages the 95 per cent limits for the Labour Party are 31.0 per cent and 37.2 per cent, while the 99 per cent limits are 30.0 and 38.2 per cent. The reader may care to confirm the further results:

Conservative:	95%	33.7% and 40.1%
	99%	32.8% and 41.0%
Liberal:	95%	9.7% and 13.9%
	99%	9.0% and 14.6%

This example illustrates a problem faced by pollsters for, although in terms of point estimates the Conservatives lead Labour, the intervals are sufficiently wide that at the selected confidence levels one could not assert that the Conservative support is greater. This is one reason why predicting the Election winner would be hazardous.

Sample Size: A Cautionary Tale

We can now be more specific in relation to a problem raised earlier which investigators need to tackle. What should be the sample size for a particular survey or project? This question is frequently posed to statisticians by researchers in other disciplines, but it is not always easy to answer to the questioner's satisfaction, especially if he has no statistical background. Assuming that one is dealing with the estimation of a proportion or percentage from a large population, then the conversation might proceed along these lines:

Researcher: 'Can you advise me on how large a sample I need to take for my survey?'

Statistician: 'Certainly, but before I do you will have to answer a few questions. First, what error are you prepared to tolerate in your estimate?'

Researcher: 'I don't want any error.'

At this point the statistician has to explain why there will always be error attached to an estimate of a population proportion derived from sample data. Assuming he succeeds, the conversation continues:

R: 'An error of no more than 1 per cent will be acceptable' (i.e. 1 per cent either way on the stated percentage).

S: 'What chance [probability] are you prepared to accept that the error will, in fact, be greater than 1 per cent?'

R (getting visibly perturbed): 'None.'

The statistician now tries to indicate the nature of confidence intervals. Again, assuming he has some success, the conversation continues:

R: 'I must have 99 per cent confidence that the error will be no greater than 1 per cent.'

S: 'Do you have any idea as to the likely percentage in the population with the attribute that you are interested in?'

R: 'None at all. If I had, I wouldn't be carrying out the survey.'

The statistician now looks at the general confidence interval for proportions and considers it in the light of the replies he has received from the researcher. As we have seen the error term is $z\sqrt{p(1-p)/n}$. The researcher has indicated that this should not exceed 1 per cent, or 0.01 (considered as a proportion). This requirement is written thus:

$$z\sqrt{\frac{p(1-p)}{n}} \leqslant 0.01$$

The researcher has also said that he seeks to achieve the 99 per cent confidence level. The appropriate z-value will thus be 2.58. One, therefore, needs n to be such that

$$2.58\sqrt{\frac{p(1-p)}{n}} \leqslant 0.01$$

No approximate information is provided as to the value of p, the population proportion, but fortunately the product $p(1-p)$ does have a maximum value of $\frac{1}{4}$, when p is $\frac{1}{2}$. It would be useful to have an estimate of p, because this could be used in the inequality and would lead to a somewhat smaller n-value. Nevertheless, in the absence of such an estimate, one completes the inequality with $p = \frac{1}{2}$ and solves it for n:

$$2.58\sqrt{\frac{0.25}{n}} \leqslant 0.01$$

Squaring both sides, one obtains

$$6.656 \times \frac{0.25}{n} \leqslant 0.0001$$

Transferring n to the other side of the inequality, we get

$$6.656 \times 0.25 \leqslant 0.0001 \times n$$

or $n \geqslant 16{,}640$. The conversation resumes:

S:	'You need a sample size of 16,640 or more.'
R (paling visibly):	'Impossible; I can't afford it.'
S:	'Perhaps you could modify your requirements?'

The researcher leaves, mumbling about statisticians who do not understand his problems.

There is a simple lesson to be learned from this little tale. With large populations, one cannot have precision without taking large samples. The resultant fieldwork will usually be quite costly both in labour and money. Before starting a survey it is necessary to specify, at any rate approximately, not just the kind of error which is tolerable but also the associated confidence level. The requirements must not be too exacting. In proceeding further, information from other investigators can be valuable, for, among other things, it may provide prior estimates for proportions which one seeks to determine. Then the appropriate sample size can be calculated and evaluated in relation to available time and finance. Where one is estimating population means, the problem of deciding sample size is generally greater since the standard error of the mean $\sigma/\sqrt{n}$ does not have a maximum value for a given n as $\sqrt{p(1-p)/n}$ does. In that case, in the absence of information from other investigators, the best procedure is probably to conduct a small pilot survey to obtain an estimate of σ from which to calculate the minimum size of sample likely to be compatible with requirements of precision.

Glossary

Random method of sample selection
Simple random sampling
Sampling frame
Random numbers
Stratified random sampling
Quasi-random sampling (or systematic sampling from lists)
Sampling distribution
Central limit theorem
Standard error of the (sample) mean
Finite population correction
Confidence intervals and limits
Unadjusted standard deviation
Adjusted standard deviation
Degrees of freedom
Student's t-distribution
Standard error of the (sample) proportion

Exercises

1 A random sample of 400 council house tenants was obtained from the lists of a city council with over 10,000 tenants and the length of the 400 tenancies (in years) was noted:

Length of tenancy	*Number of tenants*
less than 1	10
1–	54
5–	72
9–	82
13–	66
17–	52
21–	40
25 and over	24
Total	400

Obtain 95 per cent confidence limits for the mean length of tenancy for all tenants in the city's council houses.

2 The school attendance records were examined for a random sample of 40 12-year-old children in an inner-city area. The number of days of absence without a medical certificate during the year was determined for each child:

10; 0; 8; 12; 3; 2; 13; 6; 9; 15; 7; 2; 0; 11, 9; 1; 5; 21; 2; 6; 16; 8; 6; 4; 18; 6; 9; 11; 14; 12; 2; 9; 1; 3; 15; 15; 8; 7; 22; 3

Calculate 99 per cent confidence limits for the mean number of such days of absence for all 12-year-olds in the area.

3 For a random sample of 250 owner-occupied houses drawn from a city, it was found that 85 houses were jointly owned by husband and wife. Obtain 95 per cent confidence limits for the proportion of houses in the city owned jointly by husband and wife.

4 It is desired to estimate the percentage of people over 60 years of age in a large city to an accuracy of ±2 per cent with 95 per cent confidence. What is the minimum size of sample which will be required if there is indirect evidence that the proportion of such people is about one in five?

Chapter 6

Hypothesis Testing with Single Samples

As we have indicated, the basic purpose of statistical inference is to enable us to say something about the characteristics of a defined population on the basis of sample values. In Chapter 5 we looked at the problem of estimation, i.e. obtaining likely values for population parameters from known sample statistics. We now move on to examine the second fundamental area of statistical inference which is *hypothesis testing*. We shall describe how statistical hypotheses are tested in the most straightforward kinds of situation encountered in the social sciences when information from one sample is available, but first we need to clarify the precise nature of the inferences involved.

Testing Statistical Hypotheses

Essentially this area of statistics is concerned with *decision-making*. Suppose we are studying the educational qualifications of workers in the textile industry and we conduct a sample survey among this group in a region of the UK. Previous research has led us to expect that textile workers will prove to be on the whole less qualified educationally than adults as a whole. We find that of our random sample of 200 textile workers, 59 possess some educational paper qualifications (basically CSE or equivalent and above). On the other hand, we know that 40 per cent of adults in the same region possess these same qualifications. Can we conclude that the percentage of educationally qualified textile workers in the region is less than that in the adult population as a whole, or would it be wiser to view the observed difference as simply a chance occurrence due to sampling? Presented with a question such as this, a decision is called for in the face of uncertainty. In fact, either of the alternatives posed *could* be true, so the best one can do is to base a judgement on the relative probabilities involved.

When making statistical decisions, it is both necessary and useful to proceed on the basis of working assumptions about the population involved. In our example the assumption would be that the percentage of educationally qualified persons among the population of textile workers is

40 per cent, and we would seek to determine whether the proportion qualified in the sample – 29.5 per cent – was sufficiently different from this figure to justify the rejection of the initial assumption. A *statistical hypothesis* is essentially a working assumption of this type which we seek to test. As this example illustrates, a statistical hypothesis must always be specified in unambiguous quantitative terms.

More formally, one can say that at the beginning of the investigation the researcher sets up two mutually exclusive hypotheses. What is called the *null hypothesis* – denoted by H_0 – specifies particular values for one or more of the population parameters. In the example H_0 specifies that if the proportion of educationally qualified persons in the population of textile workers is represented by p, then $p = 0.40$. What is referred to as the *alternative hypothesis* – denoted by H_1 – asserts that the population parameter takes some value other than that hypothesised. Thus, in our example H_1 might state that $p \neq 0.40$. In fact, the alternative hypothesis may be either directional, or non-directional. H_1 is defined as a *non-directional hypothesis*, if it simply says that the population parameter is different from the hypothesised value specified in H_0. On the other hand, if as well as doing this, it indicates the direction of the difference (i.e. says that the population parameter is *either* greater, *or* less than the value indicated in H_0), then it is a *directional hypothesis*. Indeed in our example H_1 is a directional hypothesis, for it asserts that $p < 0.40$. Having formulated H_0 and H_1, then at a later stage when the sample data have been analysed, a decision will be made whether or not to reject H_0 in favour of H_1.

The logic of statistical inference is such that the null hypothesis is never directly proved, nor is the alternative hypothesis, since by the nature of random sampling either could be true. For instance, in our example both H_0 and H_1 are compatible with a sample proportion of 0.295. However, we may decide to *reject* the null hypothesis if on the basis of it, the sample statistic belongs to a set of possible outcomes which is both extreme and sufficiently improbable. We then decide to *assert* the alternative hypothesis. The set of extreme and unlikely outcomes is specified before we make our decision and is called the *rejection region* (or 'critical region for rejection') of the test. The (small) probability under the null hypothesis of obtaining a sample statistic in the rejection region is referred to as the *level of significance* of the test and is frequently denoted by α. If the sample statistic is found not to belong to this region, then we decide *not to reject the null hypothesis*.

Whether or not we decide to reject the null hypothesis, we may be in error, that is, whichever decision we take it may be the wrong one. For instance, we may reject a null hypothesis which is in fact true. In this case it is said that a *type I error* has been made. Alternatively we may fail to reject a false null hypothesis, in which case a *type II error* has occurred. In these circumstances it is appropriate to ask how the probability of error can be minimised. The basic problem, though, is that (other things being equal) for a given sample size an attempt to decrease the likelihood of one type of error is accompanied in general by an increase in the likelihood of the other type of error. The only way to reduce both types of error is to increase the sample size – which is therefore to be recommended – but since costs are inevitably involved, this may nevertheless be impracticable.

Generally, social scientists have adopted a conservative policy. By this is meant that there is a tendency to seek to reduce the probability of claiming a positive result which is, in fact, false, i.e. rejecting a valid null hypothesis, even at the risk of failing to claim a positive result which is, in fact, true. What this amounts to is that the chance of making a type I error is kept to a low magnitude. The investigator does this by setting a relatively small level of significance α, for this is the probability of a type I error. An α of 0.05 or 0.01 (or even 0.001) is customary. If, for instance, a 0.05 (or 5 per cent) significance level is chosen in designing a test of a null hypothesis, then there are 5 chances in 100 that we would reject the hypothesis were it valid.

The Binomial Test

Our discussion so far of hypothesis testing has been somewhat abstract and the reader may at first find this area of statistics a little daunting. However, the logic of the method can be clarified by the presentation of a straightforward example which makes use of the binomial distribution (see p. 60). We shall return to the already-presented data on textile workers shortly, but let us first modify the problem by supposing that a much smaller sample has been selected.

Example. 40.0 per cent of adults possess educational paper qualifications, but among a random sample of ten textile workers only one does. It is expected that the proportion of textile workers with these qualifications will be less than 40.0 per cent, but do these figures provide firm evidence that the proportion in this occupational group is indeed less than that of adults as a whole?

If we let p represent the proportion of textile workers in the region who possess qualifications and q the proportion who do not, then the null hypothesis H_0 is that $p = 0.40$ (and hence $q = 0.60$). On the other hand, the alternative hypothesis H_1, which is directional, is that $p < 0.40$ (and hence $q > 0.60$). Let us test the null hypothesis using the significance level $\alpha = 0.05$. In the sample one textile worker is qualified. The rejection region consists of those values for the number who are qualified, which are so small that the probability associated with their occurrence under H_0 is equal to or less than $\alpha = 0.05$. The probabilities of specific outcomes under the null hypothesis are given by the binomial distribution. In general if r is the number of cases of the identified type in a total sample of n, then the probability of this particular occurrence $Pr(r)$ is given by formula 4.13:

$$Pr(r) = \binom{n}{r} p^r q^{n-r}$$

It follows that the critical region for rejection may be determined by adding the binomial probabilities under H_0 associated with obtaining no qualified person, one qualified person, two qualified persons, etc., until the sum is greater than the significance level. The critical region consists of the largest number of extreme outcomes which is such that the associated sum is less than or equal to α. In practice, faced with a particular outcome, we do not need to determine the critical region precisely. It is sufficient to sum the probability of the particular outcome and any others more extreme, and if this sum is less than α, one can conclude that the outcome falls in the critical region.

Since $n = 10$, the probability under the null hypothesis of obtaining just one person with qualifications is

$$Pr(1) = \binom{10}{1} (0.40)^1 (0.60)^{10-1}$$

$$= \frac{10!}{9!\,1!} (0.40)(0.60)^9 = 0.0403$$

We also need to take into account the probabilities of any more extreme outcomes in the predicted direction. The only outcome more extreme than that actually obtained would have been a sample with no qualified person. Under the null hypothesis the probability of this outcome is

$$Pr(0) = (0.60)^{10} = 0.0060$$

So the probability of finding one qualified person or fewer in the sample of 10 given that 40 per cent of the population are qualified is

$$Pr(1) + Pr(0) = 0.046$$

Since this probability is less than the selected significance level of 0.05, the obtained outcome falls within the critical region and our *decision* is to reject H_0 and assert the alternative H_1. We conclude that there is evidence that the proportion of textile workers possessing educational qualifications is less than that among adults as a whole.

This illustration of the *binomial test* highlights two key points about significance tests. First, our decision may be the wrong one. We can see this because, by our own argument, although under the null hypothesis the observed value or one more extreme is unlikely, that type of outcome will in fact occur in approximately 46 cases (or samples) out of 1,000. This point carries the implication that we should treat our conclusion as *provisional* and subject to modification should further (incompatible) evidence be forthcoming. A second basic point concerns the importance that should or should not be attached to obtaining a statistically significant result. Had we found two qualified persons in the sample instead of one, then the null hypothesis would not have been rejected (since $Pr(2) = 0.1209$ and $Pr(2) + Pr(1) + Pr(0) = 0.167$). However, with such a small sample this could easily have occurred by chance whether the null hypothesis were valid or not. Hence, it can be appreciated that the line between significance and non-significance – particularly with small samples – is a thin one which is easily crossed. Thus, one has to guard against the tendency which might otherwise prevail of treating statistical significance as 'the be all and end all' of quantitative analysis and ignoring results which fall just short of significance.

The Use of the Poisson Distribution

In Chapter 5 we saw how in certain circumstances the binomial distribution may be usefully approximated by the *Poisson distribution*. This is the case when the sample size n is large and the probability p of the (rare) event occurring is very small (and subject to the condition $np < 5$). We can further exemplify statistical hypothesis testing by showing how the Poisson approximation can be used to handle an example which is again a slight modification of our original problem.

Example. Suppose it is known that 0.8 per cent of adults in a region possess a degree but that among a random sample of 500 textile workers in the region only two have degrees. It has been predicted that degree-level qualifications will be particularly rare among this occupational group, but do these figures provide evidence that the percentage with degrees among textile workers is indeed less than that of adults as a whole?

We have:

$$H_0 : p = 0.008 \qquad H_1 : p < 0.008$$

We can again (if we so choose) select a significance level $\alpha = 0.05$. The rejection region consists of low values for the number in the sample with degrees which together have a probability under H_0 equal to or less than 0.05. Now the basic result which we need to use is that the probabilities associated with specific outcomes under the null hypothesis are given by the Poisson probabilities of formula 4.15:

$$Pr(r) = \frac{a^r e^{-a}}{r!}$$

where r is the number of cases of an identifiable type in a total sample of size n and $a = np$. Since $n = 500$ and $p = 0.008$, $a = 4$. Therefore, the probability of obtaining exactly 2 workers with degrees is

$$Pr(2) = \frac{4^2 e^{-4}}{2!} = \frac{16e^{-4}}{2}$$

Similarly

$$Pr(1) = \frac{4^1 e^{-4}}{1!} = 4e^{-4} \text{ and } Pr(0) = \frac{4^0 e^{-4}}{0!} = e^{-4}$$

So the probability of the obtained outcome or one more extreme is given by

$$Pr(2) + Pr(1) + Pr(0) = \left(\frac{16}{2} + 4 + 1\right) e^{-4} = 0.238$$

Since this probability is greater than 0.05, we do not reject H_0 and instead conclude that there is insufficient evidence to assert that the percentage of textile workers with degrees is lower than that of adults as a whole.

Tests involving the Normal Distribution

So far our examples of significance tests have involved the calculation under the null hypothesis of the probabilities of certain identifiable outcomes which were then summed. We then determined whether or not the sum was less than the selected significance level. However, in many classes of problems we can use a rather more general method, which is to determine the relevant probabilities by reference to areas under the normal curve. We have already seen that there are circumstances in which the binomial distribution can be approximated by the normal distribution (p. 63), and we noted in Chapter 5 that the sampling distribution of several key statistics was normal or approximately normal. These facts can be used as the basis for many tests of statistical hypotheses. We shall consider some examples below (including our initially stated problem concerning the educational qualifications of textile workers), but we first need to examine normal curve methods against the background of our general discussion of hypothesis testing.

Suppose, given a specific null hypothesis, that the sampling distribution of a statistic y is a normal distribution with mean μ and variance σ^2. Then the distribution of $z = (y - \mu)/\sigma$ is the standard normal distribution with mean 0 and standard deviation 1. As shown in Figure 6.1, there is a probability of 0.95 that the value of z for an actual statistic y will lie between -1.96 and $+1.96$, since the area under the normal curve between these values is 0.95. However, if on choosing a single sample at random, we find that the value of z for the statistic lies outside the range -1.96 to $+1.96$, we would say that the probability of such an event occurring was only 0.05 (represented by the shaded area in Figure 6.1). It is when we obtain z-values of this magnitude – which one may characterise as extreme in that they are remote from the mean of the distribution – that we are led in the performance of the statistical test to reject the null hypothesis.

0.05 is the level of significance of the test. It represents the probability of rejecting a null hypothesis which is, in fact, true (i.e. the probability of a type I error). When we reject the null hypothesis, we say we do so at the 0.05 or 5 per cent level of significance. The set of values of z outside the range from -1.96 to $+1.96$ constitutes the rejection region of the test.

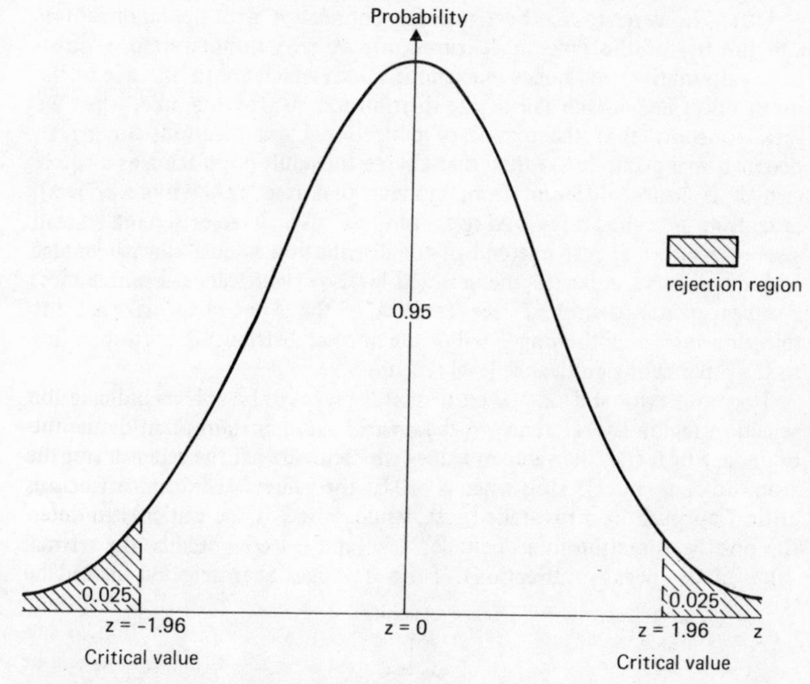

Figure 6.1 *Standard normal distribution and rejection region for two-tailed test.*

Further to the above discussion, we can formulate the following decision rule regarding the test of a statistical hypothesis:

(a) Reject the null hypothesis at the 5 per cent level of significance, if the z-value of the statistic lies outside the range from -1.96 to $+1.96$.

(b) Otherwise do not reject the null hypothesis.

Note: Since we are always dealing with probabilities in statistical decision-making, one should beware of exclusively emphasising rejection or non-rejection of the null hypothesis, but rather stress the possibility of making an error and the essentially provisional nature of the decision.

Two-tailed and One-tailed Tests

In this account of methods based on the use of the normal distribution we have so far been essentially concerned with testing H_0 against a non-directional alternative hypothesis H_1. This is apparent because we have been equally interested in extreme y-values and corresponding z-values on both sides of the mean, i.e. in both tails of the distribution. For this reason such tests are called *two-tailed tests*.

Often, however (as has been noted in connection with the binomial test and the use of the Poisson distribution), we may be interested in directional alternatives and hence in extreme values which are to one side of the mean only; i.e. in one tail of the distribution, as, for example, when we seek to show that the extent of educational qualifications among an occupational group is *less* than that among the adult population as a whole (which is clearly different from showing that it is either more *or* less). Such tests are called *one-tailed tests*. In these cases the rejection region consists of z-values at just one end of the distribution associated with an area under the curve equal to the selected level of significance. For instance, z-values greater than 1.65 (see Table A in the Appendix) make up the rejection region in the upper tail of the normal distribution corresponding to the 5 per cent significance level (Figure 6.2).

For both two- and one-tailed tests it is useful to be able to indicate the rejection region by reference to the *critical value* or *values* that define the boundary of it (i.e. the value or values which divide off the rejection region from other z-values). Thus when $\alpha = 0.05$ the values -1.96 and $+1.96$ are critical z-values in a two-tailed test, while $+1.65$ is the critical z-value in the positive direction in a one-tailed test (and -1.65 would be the critical value in the negative direction). If $\alpha = 0.01$ had been selected instead of

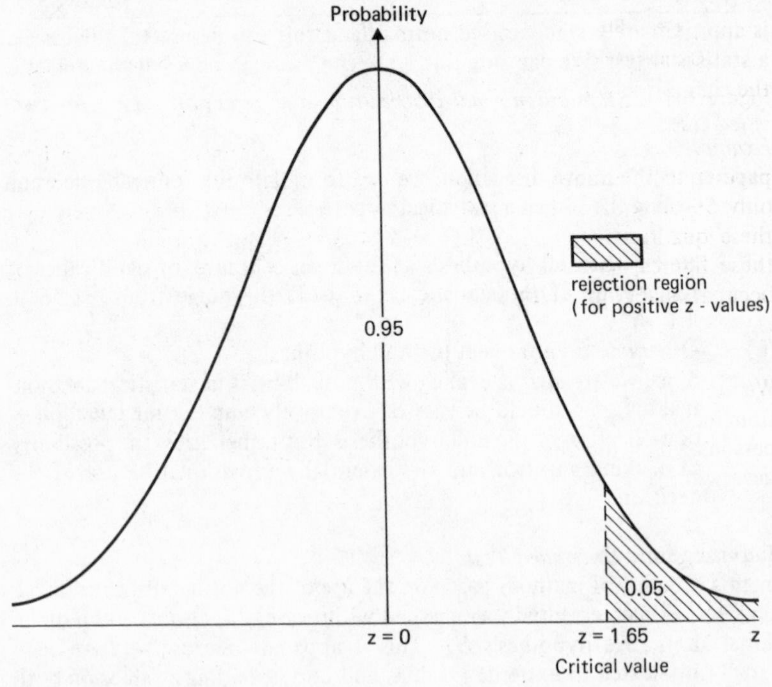

Figure 6.2 *Standard normal distribution and rejection region for one-tailed test.*

$\alpha = 0.05$, the reasoning would be precisely parallel. It can be seen from Table A that the critical z-values in a two-tailed test would then be -2.58 and $+2.58$, since 1 per cent of the area under the curve lies outside these bounds. On the other hand, Table A also reveals that in a one-tailed test with $\alpha = 0.01$ the critical z-value in the positive direction is $+2.33$ (and the negative direction -2.33).

Tests on Proportions

These results involving the normal curve can be put to immediate use in problems involving proportions. Suppose that we have a large population in which there is a proportion p of cases of a particular type (and as usual $1-p$ is designated by q). In Chapter 5 we noted the general result that if r is the number of cases of this type in a random sample of size n drawn from the population, then the sampling distribution of r/n is for fairly large n approximately normal with mean p and variance pq/n (the quantity $\sqrt{pq/n}$ being known as the standard error of the proportion). It follows that the statistic

$$\frac{r/n - p}{\sqrt{pq/n}} \tag{6.1}$$

is approximately standardised normal, and this can be used as the basis of a statistical test. We can now return to the problem posed at the outset of the chapter, which we briefly restate.

Example. It is known that 40.0 per cent of adults possess educational paper qualifications, but among a random sample of 200 textile workers only 59 do so. It is expected that the proportion of textile workers with these qualifications will be less than 40.0 per cent, but on the basis of these figures, have we good reason to believe that the proportion in this occupational group is less than that of adults as a whole? We have:

$$H_0 : p = 0.40 \quad \text{(and hence } q = 0.60) \qquad H_1 : p < 0.40$$

We are not told the significance level, but let us select $\alpha = 0.05$. The rejection region consists of low values for the sample proportion of qualified persons which together have a probability under H_0 equal to 0.05. We also have:

$$n = 200 \quad r = 59 \quad r/n = 0.295$$

and using formula 6.1, we obtain the statistic

$$\frac{0.295 - 0.400}{\sqrt{(0.40 \times 0.60/200)}} = \frac{-0.105}{0.0346} = -3.03$$

The critical value with $\alpha = 0.05$ is $z = -1.65$, since the alternative hypothesis predicts a lower value of p and hence a negative z-value. Since the obtained z-value is substantially less than the critical value, we reject the null hypothesis. We can also see that the null hypothesis would also have been rejected with $\alpha = 0.01$ (when the critical value would be -2.33), so the result is highly significant. We conclude that there is good reason to believe that the proportion of textile workers with educational qualifications is less than that of the adult population as a whole.

The calculations in the example are based on the justified assumption that the sample is small relative to the population. When the sample is more than 5 per cent of the population size, it is necessary to modify the formula for the standard error of the proportion by applying the finite population correction (see p. 83). In the notation used earlier and with the population size represented by N, the formula for the standard error of the proportion instead of being simply $\sqrt{pq/n}$ becomes $\sqrt{pq/n} \times \sqrt{(N-n)/(N-1)}$. With this modification the test on proportions would then proceed as in our example.

At this stage we can usefully take stock of the three statistical tests so far discussed in this chapter and note that between them they are sufficient to permit one to handle all the main types of problems that require the computation of binomial probabilities (as is illustrated by our presentation of examples which are essentially variations on a single theme). If n is small, then one uses the binomial test directly because the necessary computations are then easy. On the other hand, if n is large and the probability p of the event is either very small or very near unity, the Poisson approximation may be used.* Finally if n is large and p is neither small nor near one, the normal approximation may be used.

We have not so far in this chapter explicitly raised the issue of the level of scaling needed for each test – partly because our main purpose has been to concentrate on the logic of hypothesis testing itself. However, it is apparent that since binomial probabilities (and also proportions) have been involved, no more than *nominal scaling* is required for each of our three tests. This has an important bearing upon why this group of tests is so useful – and indeed popular – in the social sciences, for in many kinds of analysis only classification of attributes is possible. However as we pass on to consider two important tests on arithmetic means, we must note that they should only be used when an *interval level of scaling* has been attained.

Tests on Means

In Chapter 5 (p. 73) we considered the sampling distribution of the mean and drew particular attention to the result that for samples of size n drawn from a large intervally scaled population with mean μ and variance σ^2, the

* When p is near unity, it follows that q (i.e. $1-p$) is small and hence these two quantities may be transposed in the calculation.

sampling distribution of the mean is approximated by the normal distribution with mean μ and variance σ^2/n. For most populations likely to be encountered the approximation is close for $n > 50$. The standard deviation of the sampling distribution $\sigma_{\bar{x}}$ is generally referred to as the standard error of the mean. This general result provides the basis for a straightforward statistical test on means, because we can use the fact that for large samples

$$\frac{\bar{x} - \mu}{\sigma/\sqrt{n}} \tag{6.2}$$

is approximately standardised normal.

Example. Suppose that the mean weekly family income in a large city is known to be £115.16 with a standard deviation of £26.52. A local newspaper has published the findings of what it claims to be a survey of the income and expenditure patterns of a random sample of 100 families in the city. Among other things it is stated that the mean weekly family income of the sample was £129.40. However, it is suspected that the sample was not random (e.g. an unconfirmed report has suggested that all the respondents were home telephone contacts). Is there substantive evidence to support an assertion that the sample was biased (i.e. non-random)?

Since $n = 100$ and income is an intervally scaled variable, we can use the normal approximation. The null hypothesis is that the sample was derived randomly from a population with mean $\mu = £115.16$ and the alternative hypothesis is that it was not. The obtained value of $\bar{x}$ is £129.40. We are told that the standard deviation σ of the population is £26.52. It seems wise to choose a rather small significance level, since bias is being investigated and one may be particularly disinclined to reject a valid null hypothesis. Let us select $\alpha = 0.001$ and conduct a two-tailed test. The rejection region consists of both high and low values of the sample mean and, hence, of the z-statistic. Using formula 6.2, the approximate z-value is

$$\frac{129.40 - 115.16}{26.52/\sqrt{100}} = \frac{14.24}{2.652} = 5.37$$

Since this value exceeds the critical value of 3.30 at the 0.001 significance level, we reject the null hypothesis and conclude that there is firm evidence of a biased sample.

As was noted earlier in the example on proportions the detail of our calculation is based on the justified assumption that the sample is small relative to the population. Where this is not the case and the sample size is more than 5 per cent of the population size, the formula for $\sigma_{\bar{x}}$ should be modified by the application of the finite population correction. In those circumstances if N represents the population size, $\sigma_{\bar{x}}$ becomes $(\sigma/\sqrt{n}) \times (\sqrt{(N-n)/(N-1)})$ rather than simply $\sigma/\sqrt{n}$, but otherwise the test is unchanged.

The above procedure is a useful test for non-randomness using means. In general, however, when we wish to test whether an observed sample mean $\bar{x}$ differs significantly from some supposed (or given) population mean μ, then the standard deviation of the population σ is not likely to be known. In this circumstance we may well be tempted to replace σ in formula 6.2 by the adjusted sample standard deviation s and, in fact, this procedure yields reasonably good results when n is large (say over 60), but as n becomes smaller that way of proceeding would be seriously erroneous. In fact, as explained on p. 80 *when the population is itself normal* what we are obtaining is not z, but the statistic

$$t = \frac{\bar{x} - \mu}{s/\sqrt{n}} \tag{6.3}$$

which follows Student's t-distribution with $n - 1$ degrees of freedom.

In using this valuable result, we can proceed in much the same way as we did with z-values. The t-distribution is symmetrical just as is the normal distribution, so with the help of Table B in the Appendix we can readily define rejection regions and critical values for both two- and one-tailed tests. For instance, in Table B the first entry of the column headed $\alpha = 0.025$ tells us that with one degree of freedom (i.e. $\nu = 1$) 2.5 per cent of the area under the curve lies to the right of the t-value of 12.71. Hence with $\nu = 1$, -12.71 and $+12.71$ are critical values with $\alpha = 0.05$ in a two-tailed test. If we obtain values of t outside this range, then in the statistical test we will decide to reject the null hypothesis. Further down the same column, one can see that with $\nu = 10$ the corresponding critical values are -2.23 and $+2.23$. In addition, it is observed from the last entry in the column that as the number of degrees of freedom becomes indefinitely large the critical values are -1.96 and $+1.96$ with $\alpha = 0.05$, confirming the expectation that in these circumstances the t-distribution becomes identical with the normal distribution. On the other hand, if we were concerned with critical values in a one-tailed test at the 5 per cent level, then our attention would focus on the column headed $\alpha = 0.05$. This column reveals that for 1, 10 and an infinite number of degrees of freedom the critical values at the positive end of the distribution are 6.31, 1.81 and 1.65, respectively. One can readily illustrate the application of the t-distribution to hypothesis testing.

Example. Suppose a researcher is assessing the effectiveness of police forces and he selects a random sample of 20 from the population of all police forces in the country. Each force keeps a record of the percentage of crimes successfully cleared up. Over a period of time the distribution of this variable has been shown to be approximately normal. A target has been set that the mean percentage for crimes cleared up for all forces should be 35 per cent. In the sample the mean percentage is 31.5 per cent and the standard deviation is 9.2 per cent. Is there good reason to believe that for the population of all forces the level of effectiveness is below the target figure?

Percentage of crimes cleared up is an intervally scaled variable, and we note that for the population as a whole the distribution is approximately normal. We shall proceed as if it *were* normal, since the resulting error is likely to be slight. The null hypothesis H_0 is that the mean for the population of all forces is given by $\mu = 35$ per cent. The alternative hypothesis H_1 is that $\mu < 35$ per cent. Student's t-test is used, since we are dealing with a variable for which σ is unknown. Let us select a significance level of $\alpha = 0.05$. The sampling distribution is Student's t with d.f. = 19. Since H_1 is directional and predicts a low value for t, the rejection region consists of all values of $t < -1.73$ (see column headed $\alpha = 0.050$ and row $\nu = 19$ in Table B in the Appendix). From formula 6.3, we obtain

$$t = \frac{31.5 - 35.0}{9.2/\sqrt{20}} = -1.70$$

Since the obtained t is above the critical value, we do not reject H_0, but the result is so close that further investigation is to be recommended.

It is worth noting that had we treated the calculated quantity in our example as a z- rather than a t-statistic, then we would have been led (wrongly) to reject the null hypothesis, since the quantity obtained falls below the critical z-value of -1.65 for an α of 0.05. Hence we would have made an unjustified inference, which illustrates the importance of using Student's t-distribution in this type of problem. As indicated in Chapter 5, we are able to use the latter distribution with samples larger than 30 even when the population concerned departs noticeably from normality. These are also the cases where critical t-values approximate to corresponding z-values, but the use of the former is to be recommended since they are exact.

Hypothesis Testing and Confidence Intervals

Given familiarity with both statistical hypothesis testing with single samples, and (from Chapter 5) the construction of confidence intervals, the reader may reasonably ask what precisely is the relationship between the two. Manifestly both involve random sampling and the relation between population parameters and sample statistics, but this is hardly a detailed answer. In fact, although the explicit purposes of the two methods are different, they are very closely connected.

A significance test is used to determine by reference to a sample statistic (e.g. $\bar{x}$) whether or not to reject a null hypothesis, which specifies particular values for one or more population parameters (e.g. a value for μ). This is done after a level of significance (e.g. 0.05) has been set. In our last example taking $\alpha = 0.05$, we concluded that a sample with mean $\bar{x} = 31.5$ might come from a population with mean $\mu = 35.0$. However, keeping α

constant, it can also be shown that the same sample mean *could* come from numerous other populations with various means, e.g. it might come from a population for which $\mu = 28.0$. Hence for a given α (say, 0.05), a whole range of population parameters (e.g. μ-values) might be judged to have produced the specific sample statistic. To be precise, one can say that the $100(1-\alpha)$ per cent confidence interval contains those values of the parameter which would not be rejected at the significance level 100α per cent.

It follows that even though problems arise which call directly for the performance of a statistical test, it is useful where possible to specify in addition a confidence interval with its associated probability. Simply to compute (say) a z or t statistic and then state whether or not it is significant at level α gives less information than a confidence interval alone, since the test result may be inferred from whether or not the hypothesised value falls within the confidence interval.

Glossary

Statistical hypothesis
Null hypothesis
Alternative hypothesis
Directional hypothesis
Non-directional hypothesis
One-tailed test
Two-tailed test
Rejection region
Level of significance
Critical values
Type I error
Type II error

Exercises

1. In a particular community 10 per cent of the adult population has been classified as upper class. On the other hand, a study of the six local formal associations reveals that of their six presidents four are upper class. It is claimed that this is purely a chance phenomenon. Do you agree?
2. Support for a political party in a particular constituency is known to be at the low level of 1.6 per cent. However, it is believed that the party has even fewer supporters among skilled workers than among other groups. A random sample of 300 skilled workers proves to include two party supporters. Does this provide reliable evidence that support for the party among this occupational category is less than that among adults as a whole?
3. Records show that 31.2 per cent of adults in a city possess a driving licence, but among a random sample of 300 clerical workers as many as 114 have one. It has been predicted that the percentage possessing a licence will be relatively high among this group, but have we good reason to believe that the percentage of clerical workers in the city with licences is, indeed, higher than that of adults as a whole?
4. It is known that on an attainment test the scores for all the 10-year-old children in a region of the UK have a mean of 110 and a standard deviation of 22. For 100 10-year-olds of the region the mean score on the

test was calculated to be 116. Does it seem likely that this sample was randomly drawn from the above population? Use the 0.01 significance level.

5 In an inquiry covering a random sample of sixteen households the mean weekly expenditure on motoring was £7.56 with a standard deviation of £1.12. Is this consistent with the hypothesis that the mean monthly expenditure for the population is £8.00? Over a period of time this variable has been shown to be approximately normally distributed.

Chapter 7

Statistical Inference with Two Samples

In the two previous chapters, we considered statistical problems involving single samples, but it is perhaps even more common in the social field to encounter situations where we need to make comparisons between two or more samples. This is often done in order to establish whether statistical relationships exist between attributes and/or variables. For instance, we might seek to discover whether children from single-sex schools differ on average in their performance in attainment tests from children in co-educational schools; or, we might wish to know whether paroled prisoners have a lower rate of reconviction than comparable unparoled groups. In this chapter we deal with two sample problems, leaving for consideration in Chapters 8 and 9 procedures which may be used with larger numbers of samples.

When we have two samples, a basic statistical question is whether or not they might have been drawn from the same population or two identically distributed populations. However, the precise problem at issue varies depending upon the research design and the type of data, so we shall consider in turn in this chapter several of the more commonly encountered situations. We often ask, for instance, whether two independent random samples may have been derived from populations which differ in their average values. When the data are intervally scaled, we need to test whether there is a *difference in the arithmetic means* of the two populations. On the other hand, where we have nominal scaling, we would compare two samples in respect of the proportion of cases falling into a particular category and determine whether there was evidence of a *difference of proportions* in the underlying populations. There is, thirdly, a need for procedures directly comparable to tests involving differences of means and proportions which can be used with ordinal scaling. These latter methods are particularly useful in the social sciences, since one so often encounters ranked data, and they can also be usefully employed even with some interval level data – for instance, when the rather strict conditions needed for the application of the difference of means test (noted below) fail to apply.

Differences between Means

In the single-sample case, we required the sampling distribution of the arithmetic mean. In order to compare two sample means, we need the *sampling distribution of the difference between means*. The latter notion may initially seem complex but the underlying ideas in each case are essentially the same. Earlier it was noted that the means of samples drawn from the same population can be expected to vary. To see just how they varied, one imagined drawing an indefinitely large number of random samples of fixed size from a population and then specified the shape of the resultant distribution of sample means. Similarly, one can imagine selecting, randomly and independently, many pairs of samples – one from each of two populations – calculating the differences between the means of each pair, and determining the shape of the distribution of these differences. As the number of pairs of samples increases indefinitely, we obtain the sampling distribution of the difference between means.

A basic result which concerns this distribution can be stated as follows. If two independent random samples of size n_1 and n_2, respectively, are repeatedly drawn from normal populations, then the sampling distribution of $\bar{x}_1 - \bar{x}_2$ will be normal with mean $\mu_1 - \mu_2$ and variance $\sigma_1^2/n_1 + \sigma_2^2/n_2$, where μ_1 and μ_2 are the respective means of the populations and σ_1^2 and σ_2^2 are their variances. By 'independent' is meant that the selection of one sample in no way affects the selection of the other. As in the single-sample case the result can be generalised by the central limit theorem (p. 73) to cover populations which are not normal. To be precise, we are able to state that for *any* (intervally scaled) populations with the specified means and variances, as n_1 and n_2 become larger the sampling distribution of $\bar{x}_1 - \bar{x}_2$ approaches normality with mean $\mu_1 - \mu_2$ and variance $\sigma_1^2/n_1 + \sigma_2^2/n_2$. The practical implication is that for large samples n_1 and n_2 (i.e. of size 30 or over) from most populations likely to be encountered, the sampling distribution of $\bar{x}_1 - \bar{x}_2$ may be approximated by a normal curve. Thus, for these samples the sampling distribution of the statistic

$$\frac{(\bar{x}_1 - \bar{x}_2) - (\mu_1 - \mu_2)}{\sqrt{\sigma_1^2/n_1 + \sigma_2^2/n_2}} \tag{7.1}$$

is approximately standardised normal, and this basic fact can be used in the testing of hypotheses. The denominator of expression 7.1 (i.e. the standard deviation of the distribution) is referred to as the *standard error of the difference between means*, denoted by $\sigma_{\bar{x}_1 - \bar{x}_2}$. A straightforward problem will illustrate how this result may be applied.

Example with large samples. An achievement test (providing scores on an interval scale) was given to two random samples of size 60 drawn from pupils educated at single-sex schools and co-educational schools, respectively. The results were as follows; and we wish to test at the 5 per cent level whether or not we can reject the null hypothesis that there is no difference between the population means:

$$n_1 = 60 \quad \bar{x}_1 = 99 \quad s_1 = 5$$

$$n_2 = 60 \quad \bar{x}_2 = 101 \quad s_2 = 4$$

Since in this example the population variances σ_1^2 and σ_2^2 are unknown (which is usually the case), we use instead the unbiased sample estimates s_1^2 and s_2^2. Hence, our estimate of $\sigma_{\bar{x}_1 - \bar{x}_2}$, the standard error of the difference between means is given by

$$\sqrt{\frac{s_1^2}{n_1} + \frac{s_2^2}{n_2}} \qquad (7.2)$$

with the vital proviso that this formula is only to be used with large samples. An approximation such as this introduces an error, but for samples larger than 50 the error is not serious. Our null hypothesis H_0 is that $\mu_1 = \mu_2$, and this is being tested against the alternative hypothesis H_1 that $\mu_1 \neq \mu_2$. Hence, to reject the null hypothesis at the 5 per cent level, the value of

$$\frac{\bar{x}_1 - \bar{x}_2}{\sqrt{s_1^2/n_1 + s_2^2/n_2}}$$

must fall outside the range from -1.96 to $+1.96$ (Table A in the Appendix). In the example, our approximation to the z-value is thus

$$\frac{99 - 101}{\sqrt{25/60 + 16/60}} = -2.42$$

which is outside the specified range, so we reject the null hypothesis. We conclude that there is evidence of a difference in the average performance of pupils from the two types of schools.

Small Samples

As we have noted the large-sample method introduces an error, although a slight one. With small samples, using that method would be quite wrong. However, in cases where we have reason to believe that the two populations are *both* normally distributed, *and* have the same variance (i.e. $\sigma_1^2 = \sigma_2^2 = \sigma^2$), there is an exact way of handling small-sample problems using Student's t-distribution. In fact, it can be shown that in these circumstances the sampling distribution of the statistic

$$t = \frac{(\bar{x}_1 - \bar{x}_2) - (\mu_1 - \mu_2)}{s\sqrt{\frac{1}{n_1} + \frac{1}{n_2}}} \quad \text{where} \quad s^2 = \frac{(n_1 - 1)s_1^2 + (n_2 - 1)s_2^2}{n_1 + n_2 - 2} \qquad (7.3)$$

is the t-distribution with $n_1 + n_2 - 2$ degrees of freedom. The quantity s^2, which is derived by pooling information from both samples, is essentially the best estimate we can make of the common population variance σ^2. As was noted in the single-sample case (p. 82), this basic result concerning the t-distribution may be used in practice with populations which are either normal, or approximate to normality. When both n_1 and n_2 are greater than 30, satisfactory results may be obtained even when there are noticeable departures from normality.

Example with small samples. In a study of voting behaviour two samples of wards were taken, one in predominantly urban areas and the other in rural areas. The wards were compared with respect to the percentage of persons voting Labour in a series of local elections, with the following results:

urban areas $n_1 = 18$ $\quad \bar{x}_1 = 55\%$ $\quad s_1 = 11\%$

rural areas $n_2 = 13$ $\quad \bar{x}_2 = 45\%$ $\quad s_2 = 12\%$

It has been observed over a number of years that this variable is approximately normally distributed in each type of area. The problem is to determine whether, on these figures, there are grounds for concluding that there is a significant difference in voting behaviour between the two types of ward. The prediction has been made that urban areas will exhibit the greater percentage Labour vote. The 5 per cent level of significance has been selected.

In this example the null hypothesis H_0 is that $\mu_1 = \mu_2$, and this is being tested against the directional alternative hypothesis H_1 that $\mu_1 > \mu_2$. From expression 7.3

$$t = \frac{(55-45)-0}{s\sqrt{\dfrac{1}{18}+\dfrac{1}{13}}} \quad \text{where} \quad s^2 = \frac{17(121)+12(144)}{18+13-2} = 130.52$$

hence, $t = 2.41$.

In passing we can note that s^2, our estimate of σ, lies between $s_1^2 = 121$ and $s_2^2 = 144$. (This is a useful computational check.) Our decision is based on the t-value with its associated number of degrees of freedom, $n_1 + n_2 - 2$, or 29. In a one-tailed test a value of 1.699 is needed with $\nu = 29$ to reject the null hypothesis at the 5 per cent level (see Table B in the Appendix) and this value has, indeed, been exceeded. Our conclusion is that there appears to be good evidence of greater percentage support for Labour in urban areas.

The t*-test and Homogeneity of Variance*

As has been noted, one of the assumptions upon which the use of the two-sample t-test is based is that both samples are drawn from populations with the same variance. This assumption – referred to as *homogeneity of variance* – must be tested using another distribution: the F-distribution. If we have two random samples, one from each of two populations with equal variances, they can be expected to have variances that differ from each other. The statistic F is defined as the ratio of two *unbiased* variance estimates:

$$F = \frac{s_c^2}{s_d^2} \qquad (7.4)$$

where s_c^2 is the larger of the two unbiased estimates of the common variance and s_d^2 is the smaller.

Under the hypothesis of homogeneity of variance, the probability that the variance estimates from a pair of samples of fixed size will produce a particular F-value which exceeds unity by a given amount becomes smaller and smaller as that amount increases. A test of significance of the disparity between two variance estimates involves calculating F by formula 7.4 and then determining whether the obtained value is too large to continue maintaining the hypothesis of homogeneity of variance. For example, in the problem considered above, we were given sample values of $s_1^2 = 121$ and $s_2^2 = 144$, both being *unbiased* estimates of the common population variance. Hence from formula 7.4, the required value of F is $144/121 = 1.19$.

As with the t-distribution to test for significance using F, we need to refer to degrees of freedom but in this case there are two such numbers to consider, since there are two estimates of the common variance. In each case the number of degrees of freedom is given by the denominator of the variance estimate, i.e. one less than the sample size (cf. p. 79). What we must do is to locate that column of Table D in the Appendix, stating the number of degrees of freedom of the larger variance estimate (ν_1) and trace it down until we find the entries corresponding to the number of degrees of freedom of the smaller estimate (ν_2). In our example the larger estimate is based on $13-1$ d.f. (i.e. $\nu_1 = 12$) and the smaller on $18-1$ d.f. ($\nu_2 = 17$), so from the two arrays of Table D (in the Appendix) the critical F-values are 2.38 at the 0.05 point and 3.46 at the 0.01 point. The actual significance levels associated with these F-values are in fact double the quoted point values, i.e. $\alpha = 0.10$ and $\alpha = 0.02$, respectively, since we are here performing a two-tailed test and the F table is so constructed that it immediately gives critical values for a one-tailed test. Since our calculated value of 1.19 is considerably less than either of these, we do not reject the hypothesis of homogeneity of variance. When the F-test is performed as a preliminary to the t-test on small samples, it is considered appropriate to proceed with the latter test in the way described above, if the obtained value of F is (as in this example) less than the critical value at the 0.01 point. On the other hand, in cases where the hypothesis of homogeneity of variance is rejected, i.e. where we conclude that $\sigma_1^2 \neq \sigma_2^2$, the t-test may still be performed, but one must modify the procedure by obtaining separate estimates of the two standard deviations and using a more complex expression for the degrees of freedom (see Blalock, 1972, pp. 226–8).

Differences between Means of Related Samples (Direct Difference Method)

Sometimes we can increase the effectiveness of our research design by – instead of comparing two independently drawn random samples – comparing sets of matched pairs. We may, for instance, compare scores obtained from experimental and control groups, where individual members of each group have been associated on the basis of relevant characteristics. Alternatively, one may use a 'before and after' research design, where the scores obtained from a sample of people (or groups) before some change has been introduced are compared individually with scores obtained from the same units afterwards. The basic idea of matching in this way is to reduce the influence of extraneous factors and, hence, make it easier to draw a substantive conclusion. If there were n cases in each sample, we might be tempted to proceed as indicated earlier, using a difference of means test on two samples of size n. However, this would be quite wrong, since the samples were not drawn independently; indeed, the opposite was the case. However, what we have is a sample of independent *pairs*, so we proceed by obtaining a difference score for each pair. We can then hypothesise that the mean of the differences in the population of pairs is zero and then use a single-sample t-test to determine whether or not the hypothesis can be rejected.

Example. Suppose a political party is interested in assessing the impact of canvassing on election results. Instead of treating all wards in a large city in a uniform way (which has been the previous practice), the party in order to make the assessment adopts a different strategy: wards are carefully matched on variables thought to be relevant to voting behaviour. Those in group I are subjected to intensive canvassing, while those in group II are not exposed to canvassing at all. The following figures indicate the percentage 'swing' towards the party in each ward in the subsequent local elections. It is predicted that the average swing in group I will be greater, but is there clear evidence of the effectiveness of canvassing?

The population differences are assumed to be normally distributed and the hypothesis to be tested is that the population mean μ_D is zero. The directional alternative hypothesis is that $\mu_D > 0$. If D_1 to D_{10} are the sample differences and $\bar{D}$ and s_D are their mean and standard deviation, respectively, then under the null hypothesis the statistic

$$t = \frac{\bar{D} - \mu_D}{s_D/\sqrt{n}} \qquad (7.5)$$

is distributed as t with $n-1$ degrees of freedom. From Table 7.1, we obtain $\bar{D} = 8.7/10 = 0.87$, and

$$s_D = \sqrt{\frac{\Sigma D^2 - (\Sigma D)^2/n}{n-1}} = \sqrt{\frac{28.01 - (8.7)^2/10}{9}} = 1.507$$

Table 7.1 *Test on Related Samples applied to Data on Percentage 'Swings' in Ten Paired Wards*

Pair Number	Group I (%)	Group II (%)	Difference (I − II)(%) D	(Difference)^2 D^2
1	2.3	1.1	1.2	1.44
2	5.4	4.6	0.8	0.64
3	9.1	7.7	1.4	1.96
4	6.0	7.9	−1.9	3.61
5	3.3	4.3	−1.0	1.00
6	0.1	−2.0	2.1	4.41
7	7.5	5.6	1.9	3.61
8	1.2	−0.1	1.3	1.69
9	1.3	1.5	−0.2	0.04
10	8.9	5.8	3.1	9.61
			8.7	28.01

from which $t = 0.87/(1.507/\sqrt{10}) = 1.825(5)$. With 9 degrees of freedom, the critical value of t at the 5 per cent level with direction predicted is 1.833 (see Table B in the Appendix). Since the obtained value is below this figure, we are unable to reject the null hypothesis. However, the two values are so close that further investigation might demonstrate that canvassing does have a small positive effect.

Differences of Proportions

The second main type of situation encountered in the social sciences involving two independently drawn random samples is one where we seek to compare sample proportions. Most often the aim is to determine whether one can safely conclude that there is evidence of a difference in proportions in the two populations from which the samples are drawn. The procedure which we shall describe – and which applies for large samples – is particularly valuable because, although it may be used in conjunction with interval-level scaling, in fact it assumes no more than nominal scaling.

Suppose we have two populations in which there are respectively p_1 and p_2 proportions of cases of a particular type. What we need to do is to characterise the sampling distribution of the difference $r_1/n_1 - r_2/n_2$, where r_1 and r_2 are the number of cases in independent random samples of size n_1 and n_2 drawn from the two populations. We have already seen that the sampling distribution of r_1/n_1 is for fairly large n_1 approximately normal with mean p_1 and variance p_1q_1/n_1 (where $q_1 = 1 - p_1$) and r_2/n_2 has a corresponding distribution. In the two-sample case the required general result is that for large n_1 and n_2 the sampling distribution of $r_1/n_1 - r_2/n_2$ is approximately normal with mean $p_1 - p_2$ and variance $p_1q_1/n_1 + p_2q_2/n_2$. (This result parallels that stated for differences of arithmetic means for it can be seen that the required mean is given by the *difference* of the means of the separate sampling distributions, while the variance is given by the *sum* of the separate variances.) The standard deviation of the sampling distribution of $r_1/n_1 - r_2/n_2$ is referred to as the *standard error of the difference between sample proportions*. So we can say that for large samples, the sampling distribution of the statistic

$$\frac{(r_1/n_1 - r_2/n_2) - (p_1 - p_2)}{\sqrt{p_1 q_1/n_1 + p_2 q_2/n_2}} \tag{7.6}$$

is approximately standardised normal. The normal approximation will usually be satisfactory in applications, if r_1 exceeds 5 when $r_1/n_1 \leqslant \frac{1}{2}$ and $(n_1 - r_1)$ exceeds 5 when $r_1/n_1 > \frac{1}{2}$, and a corresponding condition holds for r_2. Let us see how we can use this result in practice.

Example (fictitious data). A random sample of 131 ex-prisoners who were paroled is taken and over a specified period of time 59 are reconvicted. On the other hand, from a random sample of 175 unparoled ex-prisoners (who had committed similar kinds of offences) 104 are reconvicted. Can one conclude that – as might be predicted – the reconviction rate is lower among those who are paroled?

The null hypothesis is that in the populations of ex-prisoners reconviction is as common among the paroled as among the unparoled, i.e. $p_1 = p_2 = p$, and we are testing this against the alternative $p_2 > p_1$. We decide to select the 1 per cent significance level partly because our conclusion may have important policy implications. We have:

$$n_1 = 131 \qquad r_1 = 59 \qquad r_1/n_1 = 59/131 = 0.450$$

$$n_2 = 175 \qquad r_2 = 104 \qquad r_2/n_2 = 104/175 = 0.594$$

and from expression 7.6, we need to determine the quantity

$$\frac{(r_1/n_1 - r_2/n_2) - 0}{\sqrt{pq(1/n_1 + 1/n_2)}}$$

Since in this example the population proportion p is unknown (as will usually be the case), we need to determine the best estimate of it. In this connection it would be quite wrong to average r_1/n_1 and r_2/n_2, since they are not based on the same number of cases. In fact in this instance n_2 is larger than n_1, so under the null hypothesis r_2/n_2 provides a more precise estimate than r_1/n_1. A little reflection indicates (what can be shown to be) the best procedure, which is to use the information from both samples by combining them and working out the proportion in the combined sample. So our best estimate of p is given by

$$\frac{r_1 + r_2}{n_1 + n_2} \tag{7.7}$$

In this case the estimate is

$$\frac{59 + 104}{175 + 131} = 0.533$$

and our approximation to the value of z is

$$\frac{0.450 - 0.594}{\sqrt{0.533 \times 0.467(1/131 + 1/175)}} = -2.50$$

Our obtained value is negative, because the reconviction rate is greater in the second sample than in the first. Given also that the calculated value is less than -2.33, the critical value of z at the 1 per cent level in a one-tailed test (Table A in the Appendix), we reject the null hypothesis. The conclusion is that we have soundly based evidence that the rate of reconviction is higher among unparoled ex-prisoners.

It is worth knowing that there are alternatives which can be used instead of the difference of proportions test. The most popular is the chi-square test which we discuss in Chapter 9. A further point to note is that the conditions required for the difference of proportions test may not be met. This may well be so in small-sample problems, in which case the Fisher exact probability test (p. 137) is a better choice.

Two Samples with Ordinal Scaling

Finally in our consideration of two-sample situations, we present tests used when data are on ordinal scales. In sociology cases are often encountered where two samples of people under investigation have been distributed into a number of ranked social classes or status groups; or data may have been derived from two samples of individuals who have themselves been asked to rank or rate items. The tests which we shall describe are even more useful than may initially appear, since they can also be employed when the rather strict conditions required for the *t*-test for small samples fail to apply (or are suspected not to).

The two tests which follow – and also some which we shall consider in Chapters 8 and 9 – are often described as *non-parametric* or *distribution-free tests*, because they do not require the population distributions to be normal, nor indeed to take any other exactly specifiable form. By contrast the *t*-test for small samples *is* based on that kind of assumption and belongs to the category of *parametric tests*. The first test which we describe is a direct alternative to the *t*-test.

The Mann-Whitney U*-test*

This is one of the most effective non-parametric tests, for it uses all the information which is provided on ranking. Suppose that we have two random samples A and B of size n_1 and n_2, respectively. The null hypothesis to be tested is that they come from populations with the same distribution. The alternative hypothesis is that the two population distributions are different. What we do is to combine the $n_1 + n_2$ cases and rank them through from the smallest to the greatest. We then focus on U_{AB} the number of times each member of A precedes a member of B, and U_{BA} the corresponding number of occasions when members of B precede those of A. Under the null hypothesis U_{AB} and U_{BA} may be expected to be approximately equal. However, if the null hypothesis is false, we would expect these two quantities to be more unequal. The statistic U employed in the test is defined as the smaller of U_{AB} and U_{BA}.

Example. Suppose that two random samples A and B of firms have been selected with seven and nine members, respectively. Those in A pay workers on an individual piecework basis, while those in B do so on a group piecework basis. All 16 firms have also been ranked according to their recent productivity performances (with rank 1 indicating the best performance). Can we reject at the 5 per cent level the null hypothesis that there is no relationship between productivity and payment method? The following is the rank order with those firms in B underlined:

$$\underline{1}\quad \underline{2}\quad \underline{3}\quad 4\quad 5\quad \underline{6}\quad \underline{7}\quad \underline{8}\quad 9\quad \underline{10}\quad \underline{11}\quad 12\quad \underline{13}\quad 14\quad 15\quad 16$$

The initial step is to determine U_{AB} by counting how many B-members follow each A-member in the rank order. The first A (number 4 in the order) has six Bs following it, as does the second A (number 5). The third A (9) precedes three Bs, while the fourth (12) precedes one. The remaining As do not precede any Bs. Thus $U_{AB} = 6 + 6 + 3 + 1 = 16$. Similarly, one can calculate that $U_{BA} = 7 + 7 + 7 + 5 + 5 + 5 + 4 + 4 + 3 = 47$. So the required statistic U, the smaller of U_{AB} and U_{BA}, is 16. In fact, inspection of the order would have suggested that U_{AB} would equal U, since so many Bs head the list or come early in the order, while As bring up the rear. However in some instances, at least, it is safer to determine both U_{AB} and U_{BA}, and we can note as a check that

$$U_{AB} + U_{BA} = n_1 n_2 \tag{7.8}$$

which in this case is 63.

To make our decision regarding the null hypothesis, we refer directly to Table E in the Appendix, which provides 5 per cent and 1 per cent critical values of U. One can note that a full array is provided for sample sizes $n_S < n_L \leqslant 25$, where n_S is the number in the smaller sample and n_L that in the larger, while a separate array is provided for samples with equal sizes. For unequal samples (which we have in our example), 5 per cent critical values are given above the diagonal and 1 per cent values below. So we inspect that part of the table above the diagonal with $n_L = n_2 = 9$ and $n_S = n_1 = 7$. The required critical value is 12, and since our obtained value of U, 16, exceeds this figure, we do not reject the null hypothesis of no relationship between productivity performance and payment method.

Some further comments on the procedure of the test may be helpful. One important point concerns what to do about ties, i.e. situations where two or more cases cannot be distinguished in the overall order. The appropriate procedure is to assign to these cases the mean of the ranks they would have received had they been distinguishable and otherwise proceed to determine U in the usual way. (Note that the calculation will often involve $\frac{1}{2}$s, etc.) When the number of ties is large we may prefer to use as an alternative the Kolmogorov-Smirnov test described later.

A further useful point of procedure concerns situations where the ns become larger and the calculation of U becomes more tedious and more likely to lead to error. In these cases it is worth knowing that identical results are obtained by using the two formulae

$$U_{AB} = n_1 n_2 + \frac{n_1(n_1 + 1)}{2} - R_1 \qquad (7.9)$$

and

$$U_{BA} = n_1 n_2 + \frac{n_2(n_2 + 1)}{2} - R_2 \qquad (7.10)$$

where R_1 and R_2 are, respectively, the sums of ranks assigned to the samples A and B of sizes n_1 and n_2. In our example, formulae 7.9 and 7.10 give

$$U_{AB} = 7 \times 9 + \frac{7 \times 8}{2} - (4 + 5 + 9 + 12 + 14 + 15 + 16) = 16$$

and

$$U_{BA} = 7 \times 9 + \frac{9 \times 10}{2} - (1 + 2 + 3 + 6 + 7 + 8 + 10 + 11 + 13) = 47$$

again providing a U-value of 16.

When the values of n_1 and n_2 are large and exceed the values shown in Table E in the Appendix, we can make use of the fact that the sampling distribution of U_{AB} or U_{BA} becomes approximately normal with mean μ_U and variance σ_U^2 given by

$$\mu_U = \frac{n_1 n_2}{2}$$

$$\sigma_U^2 = \frac{n_1 n_2(n_1 + n_2 + 1)}{12}$$

So the direct procedure to determine significance is to evaluate either U_{AB} or U_{BA} (say U_{AB}) and calculate the statistic:

$$\frac{U_{AB} - n_1 n_2/2}{\sqrt{n_1 n_2(n_1 + n_2 + 1)/12}} \qquad (7.11)$$

We will reject the null hypothesis at the 5 per cent level, if we obtain a value outside the critical range of z from −1.96 to 1.96. (We have insufficient space here to consider how the Mann-Whitney U-test is modified when the direction of the relationship is predicted. For a detailed account see, for instance, Siegel, 1956, pp. 116–27.)

The Kolmogorov-Smirnov Two-Sample Test

A second test based on ordinal scaling, which is frequently used in the social sciences, is notable for being computationally simple. The Kolmogorov-Smirnov two-sample test (K-S test) involves, like the Mann-Whitney U-test, a comparison of two independently drawn random samples, but it is unlike that test in that it is generally used when data have been grouped into three or more large categories.

Suppose as before that we have two random samples A and B of sizes n_1 and n_2 and we seek to test the null hypothesis that they come from identically distributed populations. Again, the alternative hypothesis is that the two population distributions are different. We proceed by determining the *cumulative relative frequency distribution* (c.r.f.d.) for each sample. By the c.r.f.d. of a sample we mean that distribution which is obtained by determining for each distinguishable category the proportion of cases which is less than or equal to the corresponding value. If the null hypothesis is correct, we would expect that the two sample c.r.f.d.s would be fairly similar. On the other hand, if it is false, there may be substantial differences in the two distributions. The K-S test focuses on the statistic D, which is the maximum discrepancy between the two sample c.r.f.d.s.

Example. Two random samples of adult males have been selected, one consisting of persons who have moved house in the last five years ('migrants'), the other of those who have not moved ('non-migrants'). The men have also been allocated to five categories on the basis of their occupations, which have been ranked from high to low in social prestige. Can one conclude from the following data that there is a significant relationship between migration and occupational prestige at the 5 per cent level?

Occupational Category	*Migrants*	*Non-migrants*
Professional and managerial	5	2
Intermediate	7	2
Skilled	10	3
Partly skilled	0	7
Unskilled	2	6
Totals	24	20

To calculate D, we construct from these figures a further table (Table 7.2) along the following lines. For each sample, we show both a cumulative frequency distribution (i.e. for each occupational category the number in the sample with equal or lower occupational prestige) and a cumulative

Table 7.2 *The Kolmogorov-Smirnov Test applied to Data on Migration and Occupational Prestige*

Occupational prestige categories	*Migrants*		*Non-migrants*		*Difference between proportions of migrants and non-migrants*
	number	*proportion*	*number*	*proportion*	
Unskilled	2	0.083	6	0.300	−0.217
Partly skilled or below	2	0.083	13	0.650	−0.567
Skilled or below	12	0.500	16	0.800	−0.300
Intermediate or below	19	0.792	18	0.900	−0.108
All	24	1.000	20	1.000	–

relative frequency distribution (i.e. for each occupational category the *proportion* of the sample with equal or lower occupational prestige). Then in a final column the difference between corresponding proportions is displayed and D, the required maximum discrepancy, is determined by inspection (D is taken irrespective of sign).

To test the obtained D value of 0.567 for significance, refer to Table F in the Appendix. That table is constructed along similar lines to Table E used with the Mann-Whitney U-test. When the sample sizes are unequal, we refer to the main array with critical values at the 5 per cent level shown above the diagonal and those at the 1 per cent level below. A separate array is used for samples of equal sizes. In this case, we need to use the 5 per cent section of the main array with n_L (i.e. the larger sample size) being 24 and n_S (the smaller) being 20. If D equals or exceeds the tabulated value divided by $n_S n_L$, we reject the null hypothesis. In this case the obtained value of D, 0.567, does indeed exceed $192/(20 \times 24) = 0.400$, so we conclude that there is firm evidence of a relationship between whether a man is a migrant or not and his occupational prestige.

The remaining point to note about the K-S test is what to do with larger sample sizes than those shown in Table F in the Appendix. The procedure is simply to reject the null hypothesis at the 5 per cent level if the obtained value of D exceeds $1.358\sqrt{(n_1 + n_2)/n_1 n_2}$, and at the 1 per cent level if it exceeds $1.628\sqrt{(n_1 + n_2)/n_1 n_2}$. Thus, if $n_1 = 50$ and $n_2 = 100$ and a D value of 0.300 is obtained, then in a significance test at the 5 per cent level this latter figure is compared with $1.358\sqrt{(50 + 100)/(50 \times 100)} = 0.235$. Since the critical value is exceeded, the null hypothesis would be rejected. (In the interests of brevity and simplicity in this account of the K-S test, we have not considered problems where the direction of the relationship is predicted. For appropriate details in that case, see Siegel, 1956, pp. 127–36; Blalock, 1972, pp. 262–5.)

Conclusion

In this chapter, we have discussed five important significance tests appropriate for differing kinds of problem frequently encountered in the social sciences. With one exception – the t-test on the means of related samples – they involve a comparison of two independently drawn random samples. Besides the single example given, many other tests are available for handling 'matched pair' or related sample situations (see Siegel, 1956, ch. 5) but in our experience, though they are frequently used in experimental psychology, these tests are less often employed in the social studies as a whole. The underlying problem in research design is that one often has insufficient understanding of which variables or attributes to use for matching purposes or an inability to select appropriate cases even when one has this knowledge. Hence, in sociology and other social subjects the tendency is to rely more heavily on tests involving a comparison of independent samples.

The main problem for the statistically inexperienced is making the appropriate selection among these latter tests. In this connection the vital point is the type of data available and – in the specific case when data are intervally scaled – the normality assumption. If the conditions required for the difference of means test on small samples are met, then that is the best test to use in the sense that a false null hypothesis is thereby maximally likely to be rejected, but otherwise the use of the test may lead to misleading results. To avoid this risk the Mann-Whitney *U*-test and the Kolmogorov-Smirnov test are recommended as highly effective alternatives. Of the procedures considered in this chapter the very useful difference of proportions test is in a category of its own, for it requires only nominal scaling. For this reason, it can in some respects be categorised with – and serve as an introduction to – the other tests involving that type of scaling which we consider in Chapter 9.

Glossary

Sampling distribution of the difference between means
Standard error of the difference between means
F-distribution
Homogeneity of variance
Independent samples
Related samples
Standard error of the difference between proportions
Parametric test
Non-parametric (or distribution-free) test

Exercises

1 (a) An achievement test was given to two random samples A and B drawn from pupils educated in 'streamed' and 'unstreamed' secondary schools, respectively. Do the following results provide evidence of significant differences in average performance?

Sample A	$n_1 = 70$	$\bar{x}_1 = 88$	$s_1 = 7.1$
Sample B	$n_2 = 80$	$\bar{x}_2 = 92$	$s_2 = 8.2$

(b) A further study was conducted on two smaller random samples C and D drawn from an educationally deprived area. Experience over several years shows that achievement test scores are approximately normally distributed for pupils from each type of school in the area. What corresponding conclusion should be drawn in this case?

Sample C	$n_1 = 14$	$\bar{x}_1 = 76$	$s_1 = 4$
Sample D	$n_2 = 16$	$\bar{x}_2 = 81$	$s_2 = 5$

2 An advertising agency is interested in determining the effectiveness of some new techniques. A random sample of nine subjects is exposed to a series of films, lectures, etc. designed (and expected)

to influence favourably attitudes towards a set of products. The attitudes of the group are individually assessed before and after exposure (higher scores indicating more favourable attitudes). Do the following data provide sound evidence of the effectiveness of the techniques? State the assumptions on which your statistical test is based:

Subject	*Before*	*After*
1	23	29
2	22	20
3	29	33
4	8	11
5	20	24
6	2	7
7	13	12
8	4	16
9	14	13

3 In a random sample inquiry conducted in a large civic university 47 out of 200 students resident on campus favoured the setting up of a crèche, whereas 50 out of 300 students in lodgings and flats favoured it. Is there evidence of a real difference of opinion on this issue?

4 Nineteen clubs randomly sampled in an English county have been classified on the basis of their memberships as *either* predominantly middle class, or predominantly working class. They have also been ranked according to the degree of formality of their meetings (rank 1 indicating the greatest formality). The ranks are as follows:

Predominantly middle class: ranks 1,2,3,4,6,7,8,11,12.

Predominantly working class: ranks 5,9,10,13,14,15,16,17,18,19.

Is there a significant difference between the two groups of clubs at the 1 per cent level?

5(a) A sociologist has been able to rank occupations from high to low in social prestige, using five categories. Every member of a random sample of employed men has been asked whether or not he favours the abolition of fox hunting. Do the following results indicate a significant relationship between occupational level and opinions expressed on this issue? (Note that the two categories – favouring and opposing abolition – may be considered to be independent random samples from the populations with these views, since a random sample assures independence between selected subsamples.)

Occupational level	*Favours*	*Opposes*
Professional and managerial	2	6
Intermediate	4	7
Skilled	5	4
Partly skilled	11	6
Unskilled	2	2
Total	24	25

(b) Suppose the samples had been ten times as large but the proportional distributions had been the same (i.e. every figure in the table is multiplied by ten). What would be the answer to the corresponding question in that case?

Chapter 8

An Introduction to Analysis of Variance

In Chapter 7, we considered some statistical tests which may be used to compare the data derived from two random samples. It is natural to inquire what methods are available when we seek to go beyond this and analyse the differences between three or more samples. The most general procedure appropriate to these cases is known as *analysis of variance*.* This method is very general and can be adapted to quite complex research designs, but as an introduction we shall consider the straightforward case, which is particularly useful in the social sciences, when we aim to compare the means of a number of independently drawn random samples (with data on an interval scale). Analysis of variance is based on the idea that the total variation associated with the values of a sample can be divided into component parts in order to arrive at independent estimates of a population variance, so we must first consider how this may be done and then we can provide an illustration.

Sums of Squares and Variance Estimates

We have previously seen (p. 79) that the unbiased variance estimate determined from a random sample of size n is given by

$$s^2 = \frac{\sum_{i=1}^{n}(x_i - \bar{x})^2}{n-1} \tag{8.1}$$

where $\bar{x}$ is the sample mean. We also have an alternative form of this result which is, in fact, computationally simpler:

*If one wishes to test for differences among the means of n independent samples and the conditions for the two-sample t-test apply, then it is possible to proceed by carrying out $\binom{n}{2}$ t-tests. However, this would be cumbersome. By contrast, analysis of variance is a general method which determines for all n samples whether or not there are any significant differences.

$$s^2 = \frac{\sum_{i=1}^{n} x_i^2 - \left(\sum_{i=1}^{n} x_i\right)^2 / n}{n-1} \tag{8.2}$$

For example, if we have the following ten values:

21 26 24 20 26 18 30 18 19 25

then $\sum x_i = 227$ $\bar{x} = 227/10 = 22.7$ $\sum x_i^2 = 5{,}303$

and from expression 8.2 $s^2 = \dfrac{150.1}{9} = 16.7$

Further to formula 8.1, the numerator of this variance estimate – 150.1 – is referred to as the *total sum of squares* of deviations from the mean for the sample. The denominator $n - 1 = 9$ specifies the *number of degrees of freedom* of the estimate.

Next let us consider the effect of randomly splitting the sample of ten values into (say) three groups containing three, three and four values respectively. Suppose this generates the groups 25, 21, 18; 24, 26, 30; and 18, 19, 20, 26. Then we have here three independent random samples each of which can be used to provide an unbiased variance estimate just as the original overall sample of ten values did. Let us designate each group by using the subscripts 1, 2 and 3, respectively and understand that the summation is over all values in the group. Then in determining our three unbiased variance estimates, we have for the first group,

$n_1 = 3,$ $\sum x_1 = 64,$ $\bar{x}_1 = 21.3\dot{3},$ $\sum x_1^2 = 1{,}390$

and hence, using formula 8.2,

$$s_1^2 = \frac{24.6\dot{6}}{2}$$

for the second group,

$n_2 = 3,$ $\sum x_2 = 80,$ $\bar{x}_2 = 26.6\dot{6},$ $\sum x_2^2 = 2{,}152$

and hence,

$$s_2^2 = \frac{18.6\dot{6}}{2}$$

and for the third group,

$n_3 = 4,$ $\sum x_3 = 83,$ $\bar{x}_3 = 20.75,$ $\sum x_3^2 = 1{,}761$

and hence,

$$s_3^2 = \frac{38.75}{3}$$

Now we can use these independent variance estimates to provide a further more accurate unbiased estimate (just as we did in the two-sample t-test; see p. 107). We do this by pooling the information from the three samples. To be precise, our estimate s_w^2 is arrived at by adding the sums of squares for all three samples and dividing by the sum of the degrees of freedom:

$$s_w^2 = \frac{\Sigma(x_1 - \bar{x}_1)^2 + \Sigma(x_2 - \bar{x}_2)^2 + \Sigma(x_3 - \bar{x}_3)^2}{2 + 2 + 3}$$

However, since for the first group $\Sigma(x_1 - \bar{x}_1) = 2s_1^2$ with similar results applying for the second and third groups, we can also determine s_w^2 directly from s_1^2, s_2^2 and s_3^2:

$$s_w^2 = \frac{2s_1^2 + 2s_2^2 + 3s_3^2}{2 + 2 + 3} = \frac{24.6\dot{6} + 18.6\dot{6} + 38.75}{7}$$

$$= \frac{82.1}{7} = 11.7$$

As was true for the three individual estimates, the denominator – 7 – indicates the number of degrees of freedom of the estimate. However, what exactly is the numerator – 82.1? From the definition of s_w^2 it may be described as the total sum of squares of the values within each group about the mean of that group. It is, in fact, that amount of the total sum of squares of the original sample – 150.1 – which originates from variation within the three separate groups. The difference, namely, $150.1 - 82.1 = 68.0$, arises from the separation of the original ten scores into three groups. That separation led to groups with differing means and the variation of these latter about the overall sample mean is reflected in the estimate s^2, but not in s_w^2. The truth of this statement is illustrated, if we determine from the overall mean $\bar{x} = 22.7$ and the group means $\bar{x}_1 = 21.3\dot{3}$, $\bar{x}_2 = 26.6\dot{6}$ and $\bar{x}_3 = 20.75$, the weighted sum $n_1(\bar{x}_1 - \bar{x})^2 + n_2(\bar{x}_2 - \bar{x})^2 + n_3(\bar{x}_3 - \bar{x})^2$, giving $3(21.3\dot{3} - 22.70)^2 + 3(26.6\dot{6} - 22.70)^2 + 4(20.75 - 22.70)^2 = 68.0$, which is precisely the difference between the total sum of squares for the sample and the sum of squares derived from within the groups.

So what we have done is to divide the total sum of squares for the sample (150.1) into two parts, one originating from variation within the groups (82.1) and the other from variation of the means of the groups about the overall sample mean (68.0). Now it can be shown *for normal populations* that the latter quantity can be used to calculate an unbiased variance estimate s_b^2, which is independent of that obtained from variation within the groups (s_w^2); s_b^2 is obtained by dividing the above weighted sum by its associated number of degrees of freedom, which is simply one less than the number of groups. So s_b^2 is given by

$$s_b^2 = \frac{n_1(\bar{x}_1 - \bar{x})^2 + n_2(\bar{x}_2 - \bar{x})^2 + n_3(\bar{x}_3 - \bar{x})^2}{3 - 1}$$

$$= \frac{68.0}{2} = 34.0$$

We can now draw up the table which is conventionally displayed in analysis of variance problems (Table 8.1). It can be observed that the degrees of freedom total as well as the sums of squares. The reason for our choice of subscripts for s_w^2 and s_b^2 can be seen, for the former is a variance estimate derived from variation *within* the groups while the latter comes from variation *between* groups. In the final column the ratio of our two independent variance estimates $s_b^2/s_w^2 = 2.9$ is shown. Since we have two independent and unbiased estimates of a single population variance, we can determine whether the disparity between them is within the range we would expect using the *F*-test which we introduced in the last chapter (p. 108). As noted there, we refer to Table D in the Appendix, which is so constructed as to give us critical *F*-values in a one-tailed test. We proceed by determining whether or not the ratio s_b^2/s_w^2 exceeds the critical value at (say) the 5 per cent level shown in Table D. The critical value is displayed in the table opposite the associated number of degrees of freedom of, first, the larger estimate $s_b^2(\nu_1)$, and secondly, the smaller $s_w^2(\nu_2)$.* In this instance with $\nu_1 = 2$ and $\nu_2 = 7$, the critical value with $\alpha = 0.05$ is given as 4.74, which exceeds the obtained value. Therefore, we can confirm that the disparity between our two estimates could easily have occurred by chance.

At this point we can usefully take stock. We have shown that a single sample can be randomly divided into a number of separate groups and this can yield two independent unbiased estimates of a common population variance. These estimates can themselves be compared using the *F*-test. In our example the result was non-significant, but this was hardly surprising, since given that we had divided the initial sample randomly, we were assured that the ratio of our two estimates would conform to the pattern

Table 8.1 *Analysis of Variance Summary Table*

Source of variation	*Sums of squares*	*Degrees of freedom*	*Variance estimates*	*Variance ratio*
Between groups	68.0	2	$34.0(s_b^2)$	$2.9(s_b^2/s_w^2)$
Within groups	82.1	7	$11.7(s_w^2)$	
Total	150.1	9		

*Were $s_b^2/s_w^2 < 1$, it would not be appropriate to refer to Table D (pp. 189–90), for that table is concerned with the ratio of a larger to a smaller variance estimate. However, one would be immediately entitled to conclude that a ratio less than unity could have arisen by chance.

of the *F*-distribution. The method is quite general but what has still to be explained is how the basic result is put to effective use.

We do this by working in the reverse direction. Suppose at the outset, that we have a number of independent random samples assumed to have been taken from normal populations with the same variance. Suppose also, that we have set up the null hypothesis that they come from populations with the same mean. Then together, they effectively constitute a single random sample from one normal population (since any normal distribution is uniquely defined by its mean and variance). We can then proceed as though this larger random sample had been divided up into the separate groups and construct our analysis of variance table accordingly. If the null hypothesis is true, then we are likely to end up with a non-significant variance ratio value. However, suppose on the contrary that the null hypothesis is false and that the samples come from populations with differing means. Then what will tend to happen will be that the between-groups variance estimate s_b^2 – which reflects the differences between the group means and the overall mean – will tend to be large relative to the within-groups estimate s_w^2. In this case the variance ratio s_b^2/s_w^2 will be greater and the use of the *F*-test may well lead to the rejection of the null hypothesis. Hence, the analysis of variance procedure can form the basis of a highly effective significance test on the differences of the means of several samples, as in the following problem.

Example. In Table 8.2 the scores on an attitude test of independent random samples of secondary school children from five differing kinds of home background are displayed. The scores are intervally scaled and it is known that for large populations they tend to follow the normal distribution. Is there evidence that the mean score on the test varies depending upon type of home background?

Table 8.2 *Data on Attitudinal Scores for Analysis of Variance*

	Sample					
	1	*2*	*3*	*4*	*5*	
	17	30	25	26	29	
	26	18	25	32	41	
	24	19	33	21	29	
	20	25	19	24	32	
	25		25	18	38	
			24	27	36	
Totals	112	92	151	148	205	Grand total = 708
Number of scores	$n_1 = 5$	$n_2 = 4$	$n_3 = 6$	$n_4 = 6$	$n_5 = 6$	Total number of observations $n = 27$
Sample mean	22.4	23.0	25.2	24.7	34.2	Overall mean $\bar{x} = 26.2$
Sums of squares of scores	2,566	2,210	3,901	3,770	7,127	Total = 19,574

Our null hypothesis H_0 is that all five samples come from populations with the same mean and the alternative H_1 is that they do not. We shall proceed on the assumption that the five population variances are equal (though this can itself be tested using Bartlett's test; see, for instance, Fraser, 1958, p. 260). The five samples can, therefore, be thought of as together constituting a single overall sample from a normal population, so we can proceed to construct the analysis of variance table.

The total sum of squares (SS) is found from the numerator of formula 8.2:

$$\text{total } SS = 19{,}574 - \frac{708^2}{27} = 1{,}008.7$$

where the subtracted item is referred to as the *correction factor*. The total degrees of freedom is $(n-1) = 26$. The total SS must next be divided into two parts, one due to variation within the five samples and the other to variation of the individual sample means about the overall mean. The latter SS is the easier to calculate. Above, we obtained it by squaring the difference between each sample mean and the overall mean, multiplying each of these quantities by the associated sample size, and then summing the results. It is a little more direct to square the sample totals, divide each square by the number of scores in the sample, add the results and subtract the correction factor. We obtain

$$\text{between-samples } SS = \frac{112^2}{5} + \frac{92^2}{4} + \frac{151^2}{6} + \frac{148^2}{6} + \frac{205^2}{6} - \frac{708^2}{27} = 514.5$$

and the associated number of degrees of freedom is one less than the number of samples, i.e. 4.

The within-samples SS can then be calculated by subtracting the between samples SS from the total SS:

$$\text{within-samples } SS = 1{,}008.7 - 514.5 = 494.2$$

If we so choose, we can check this last calculation by finding the sums of squares of the individual samples (each given by the numerator of formula 8.2) and adding them thus:

$$\text{within-samples } SS = \left(2{,}566 - \frac{112^2}{5}\right) + \left(2{,}210 - \frac{92^2}{4}\right) +$$

$$\left(3{,}901 - \frac{151^2}{6}\right) + \left(3{,}770 - \frac{148^2}{6}\right) + \left(7{,}127 - \frac{205^2}{6}\right) = 494.2$$

confirming the above figure. The within-samples degrees of freedom can be obtained most directly by subtraction of the between-samples degrees of freedom from the total degrees of freedom, i.e. $26 - 4 = 22$. We can now complete the analysis of variance table by simply determining by division the two variance estimates and the variance ratio (Table 8.3).

Our final step is to compare the calculated variance ratio with the

Table 8.3 *Analysis of Variance Table for Attitudinal Scores*

Source of variation	*Sums of squares*	*Degrees of freedom*	*Variance estimates*	*Variance ratio*
Between samples	514.5	4	128.63	5.73
Within samples	494.2	22	22.46	
Total	1,008.7	26		

critical value at the 5 per cent level in the *F*-test. Since the degrees of freedom of the between-samples variance estimate is $\nu_1 = 4$ and of the within-samples estimate $\nu_2 = 22$, we can see from Table D in the Appendix that the required critical value is 2.82. As this value has been substantially exceeded, we reject H_0 and conclude that there is clear evidence that the mean score on the test does vary depending upon home background. If we choose, we can follow up this result by performing the two-sample *t*-test on pairs of samples in order to determine which home backgrounds differ significantly (Guildford, 1956, pp. 263–4). Inspection of Table 8.2 suggests that the significant *F*-value has arisen substantially because the mean of sample five is considerably greater than those of the other four samples.

The Kruskal-Wallis Test

If the requirements for analysis of variance and the associated *F*-test are not met (i.e. the normality assumption and the need for samples to be selected from populations with the same variance), an alternative procedure is available which is particularly easy to execute. This is the Kruskal-Wallis test, which, like the Mann-Whitney *U*-test and the Kolmogorov-Smirnov two-sample test described in Chapter 7, is characterised as a non-parametric test because it does not require the population distribution(s) to take any exactly specifiable form. All that is needed for the test is independent random samples with *ordinally scaled data* (and it is also assumed that the variable being considered has an underlying continuous distribution).

Example. Suppose that the UK has been divided up into three types of area: urban, small town and rural. From each type seven police districts have been randomly chosen. All 21 of the selected police districts are then ranked according to their crime rates (rank 1 indicating the lowest crime rate). Are there significant differences at the 5 per cent level in the ranks obtained from the three types of area (Table 8.4)?

The test focuses on the sum of the ranks in each column, denoted in this three-sample problem by R_1, R_2 and R_3. The aim is to determine whether these sums (or more precisely the mean rank for each column) are so different in magnitude that it is unlikely that the samples derive

Table 8.4 *The Kruskal-Wallis Test applied to Criminal Data*

	Crime Ranks		
	Urban areas	*Small-town areas*	*Rural areas*
	8	2	7
	12	14	10
	20	6	1
	4	11	15
	13	18	5
	16	9	17
	21	19	3
Totals	$R_1 = 94$	$R_2 = 79$	$R_3 = 58$

from the same population or identically distributed populations. In this case the null hypothesis H_0 is that crime rates are distributed in the same way in each type of area, and this is being tested against the alternative that they are not. We proceed by determining the statistic

$$H = \left[\frac{12}{N(N+1)} \sum_{i=1}^{k} \frac{R_i^2}{n_i}\right] - 3(N+1) \quad (8.3)$$

where k is the number of samples, n_i the number of ranks in the ith sample, N the total number of ranks in all the samples, R_i the sum of the ranks in the ith sample and $\sum_{i=1}^{k}$ indicates that one must sum over the k-samples. In our example

$$H = \frac{12}{21 \times 22}\left[\frac{94^2}{7} + \frac{79^2}{7} + \frac{58^2}{7}\right] - 3 \times 22 = 2.427$$

The distribution of H under the null hypothesis has been determined and Table G in the Appendix indicates the critical values at the 5 and 1 per cent levels. In this instance $k = 3$ and each sample size is 7, so referring to the second column of the table, we observe that an H-value of 5.819 or more must be obtained to reject the null hypothesis at the 5 per cent level. Since our calculated value is less than this, we cannot reject H_0 and our conclusion is that crime rates may well be distributed in the same way in the three types of area.

The question arises with the Kruskal-Wallis test of how to modify the procedure, if there are tied observations (e.g. in our example if two or more crime rates are identical). In this instance, one assigns to these cases the mean of the ranks they would have received had they been distinguishable. One then computes H by means of formula 8.3, *divided by a correction factor for ties* which is

$$1 - \frac{\Sigma T}{N^3 - N} \quad (8.4)$$

where $T = t^3 - t$ (t being the number of tied observations assigned the same rank) and ΣT indicates that one must sum over all tied groups. If there is one group of four tied observations $t = 4$ and hence $T = 4^3 - 4 = 60$, so if $N = 21$ as in our example above, the correction factor would be

$$1 - \frac{60}{21^3 - 21} = 0.993(5)$$

Division by this factor serves to increase (slightly) the value of H and thus makes it more likely that the null hypothesis will be rejected compared with the use of an uncorrected H-value.

Finally, we need to know how to determine significance when the sample sizes are larger than those shown in Table G in the Appendix. In this connection, one must briefly refer to the chi-square (χ^2) test considered in detail in Chapter 9. The test involves the use of a family of distributions – the chi-square distributions – which like the t-distributions depend upon the number of degrees of freedom ν. Table C (p. 188) shows the critical values of χ^2 which are exceeded with probability α for specified numbers of degrees of freedom. Under the null hypothesis for larger sample sizes the statistic H is approximately distributed as χ^2 with $\nu = k - 1$, so to determine critical values we merely need to refer to Table C. The approximation is, in fact, fairly close if there are more than five observations in each sample, so it can be shown how it would work in our original example. In that instance, we obtained an H-value of 2.427, and from Table C (second row), it is seen that with $\nu = k - 1 = 2$ the critical value at the 5 per cent level is 5.991. So this confirms the previous decision not to reject the null hypothesis.

Glossary

Analysis of variance
Total sum of squares
Between-samples sum of squares
Within-samples sum of squares
Correction factor

Exercises

1 For each of four types of area some local employment districts were randomly selected. The unemployment rate (a percentage of the labour force) was calculated for each employment district. Determine whether there are significant differences in the unemployment rates of the four types of area, using (a) the ordinary one-way analysis of variance procedure, and (b) the Kruskal-Wallis test. Compare your results:

Urban areas	*Large-town areas*	*Small-town areas*	*Rural areas*
7.60	6.69	4.38	3.96
7.02	7.69	4.97	2.38
5.10	4.40	3.90	4.03
4.46	5.72	4.97	2.98
5.18	5.62	3.98	2.44
6.93	4.99	3.56	
	5.60	3.88	

2 In a study of religiosity a group of ten subjects is selected randomly from each of six Protestant sects. The individuals are scored according to the extent of their religious involvement (higher scores indicating greater religiosity). Previous research has suggested that for large populations the distribution of scores follows the normal distribution fairly closely. Have we evidence that the sects differ in their mean religiosity scores?

Sects

1	*2*	*3*	*4*	*5*	*6*
51	50	46	24	51	65
50	46	43	48	45	46
49	50	39	61	62	51
51	44	40	51	51	46
49	44	55	46	44	58
51	49	49	46	20	46
46	46	43	46	54	52
45	44	42	42	36	52
45	51	44	48	64	44
47	47	44	46	40	50

Chapter 9

The Chi-Square Test and Contingency Problems with Nominal Scales

In this chapter, we consider some statistical tests and measures which may be used to analyse the relationship between two attributes. These procedures are particularly valuable in the social field, for we frequently encounter populations which can be divided up according to such attributes as sex, marital status and religion. As has been indicated, tests are available for drawing inferences when a single attribute is involved, but the more common situation in the social sciences is where we seek to examine the inter-relationship between two or more categorisations. When members of a sample are classified simultaneously according to two nominal scales, a rectangular array of frequencies is produced and what is known as a *contingency table* is displayed (e.g. Table 9.1).* In these cases, we may need to determine not simply *whether* a statistical relationship holds between attributes, but also what the *strength* of that relationship is. In this chapter these two issues are examined successively.

The Chi-Square Test

Many problems concerning relationships between attributes can be dealt with by the chi-square (χ^2) test. We may want to answer questions such as: Is there variation in the child-rearing patterns of those belonging to different social classes? Is there a statistical relationship between political opinion and religious affiliation? The use of the test is effectively displayed by a simple example, since the extension in procedure to more elaborate cases is quite straightforward. Suppose that we aim to provide evidence in respect of the second question, and to this end we have selected a random sample of 600 people from the adult population of a city and gathered data on political allegiances and religious affiliation which can be summarised in a 3×3 contingency table (Table 9.1).†

*Sometimes for these purposes data on an ordinal or even an interval scale are grouped and treated as if nominal. However, this can involve the discarding of information and is not always to be recommended.

†In the interests of simplicity of exposition the minority groups, 'don't knows', etc., which would be encountered in practice have been eliminated from Table 9.1.

Table 9.1 *Religious Denomination and Party Allegiance (percentages in brackets)*

Denomination	*Conservative*	*Political Party* *Labour*	*Liberal*	*Totals*
Church of England	153 (49.4)	98 (31.6)	59 (19.0)	310 (100.0)
Roman Catholic	53 (35.8)	75 (50.7)	20 (13.5)	148 (100.0)
Free Church	45 (31.7)	68 (47.9)	29 (20.4)	142 (100.0)
Totals	251	241	108	Grand Total 600

Inspection of the table can be very informative, especially if we determine percentages, for we can note among other things that 49.4 per cent of Anglicans are Conservatives, while the corresponding figure for Roman Catholics is only 35.8 per cent and of Free Church members 31.7 per cent. Generally speaking, there does seem to be an observable tendency for the Conservative Party to gain relatively more support from Anglicans, while the Labour Party does rather better among members of the two other religious groups. However, are these (and other) observable differences statistically significant or simply a feature of this particular random sample?

In general the chi-square test involves a comparison between an *observed* number of cases falling into each category and an *expected* number of cases on the basis of a theoretical distribution or a hypothesis to be tested. In the case of the above type of contingency table no theoretical distribution is available to be used to determine expected frequency values, so these latter values are determined from the observed frequency values themselves. In the example one takes as the statistical hypothesis to be tested, that in the population as a whole the distribution of political allegiances is the same among those of differing denominations. Given this working assumption and also the fact that the sample was randomly selected, an expected set of frequencies can then be computed from the marginal (i.e. column and row) totals and the grand total.

One reasons as follows. Consider the expected number of Anglicans who are Conservatives E_{11} (the subscripts indicating the first row and first column). If there is no association between religious denomination and politics, then the proportion of Anglicans who are Conservatives should be the same as the proportion of Conservatives in the sample as a whole. Thus:

$$\frac{E_{11}}{310} = \frac{251}{600} \quad \text{or} \quad E_{11} = \frac{310 \times 251}{600} = 129.7$$

By similar reasoning the expected number of those in the Church of England who support Labour, E_{12}, should be given by

Table 9.2 *Religion and Politics: Expected Frequencies*

	Conservative	*Labour*	*Liberal*	*Totals*
Church of England	129.7	124.5	E_{13}	310
Roman Catholic	61.9	59.4	E_{23}	148
Free Church	E_{31}	E_{32}	E_{33}	142
Totals	251	241	108	Grand Total 600

$$E_{12} = \frac{310 \times 241}{600} = 124.5$$

Turning to the second row, the expected number of Roman Catholics who are respectively Conservatives (E_{21}) and Labour supporters (E_{22}) can be calculated:

$$E_{21} = \frac{148 \times 251}{600} = 61.9 \qquad E_{22} = \frac{148 \times 241}{600} = 59.4$$

At this point the table of expected frequencies takes the form shown in Table 9.2.

Although the other five expected values could be calculated in the same way, it is clear that they can be determined by subtraction from the marginal (column and row) totals:

$$E_{13} = 310 - 129.7 - 124.5 = 55.8$$

$$E_{31} = 251 - 129.7 - 61.9 = 59.4, \text{ etc.}$$

In fact, once E_{11}, E_{12}, E_{21} and E_{22} have been ascertained from the statistical hypothesis, the remaining expected values can be deduced.*

In a contingency table such as this the least number of expected values which has to be calculated directly from the hypothesis before all the others are given by subtraction is known as the *number of degrees of freedom* (ν). In general, if r is the number of rows and c the number of columns in a table, then the number of degrees of freedom is given by

$$\nu = (r-1)(c-1)$$

In the example $r=3$ and $c=3$, which confirms that $\nu=(3-1)(3-1)=4$.

The chi-square statistic essentially provides a measure of the overall difference between the observed and the expected set of values and it is a measure for which under the hypothesis of no association the sampling distribution is known. It is defined thus:

*The process of subtraction as outlined may lead to (small) errors at the last decimal place. In practice it would be more accurate to calculate all E-values directly.

$$\chi^2 = \sum \left[\frac{(O-E)^2}{E}\right] \qquad (9.1)$$

where O and E refer, respectively, to the observed and expected frequencies for each cell and it is understood that the summation is over all cells. The values of $(O-E)^2/E$ for the nine cells are

4.186	5.641	0.184
1.280	4.097	1.638
3.491	2.123	0.452

Hence, the required value of χ^2 is 23.09 (to two decimal places).

We are now ready to set the significance level and reach a decision about the statistical hypothesis of no association. Suppose that we intend to make it harder to reject the null hypothesis (i.e. be 'conservative' in the statistical sense!) then we might decide to use the 1 per cent level of significance. We refer to the chi-square distribution (Table C in the Appendix) and look across the row corresponding to four degrees of freedom (i.e. $\nu = 4$). A value of 13.28 or more would occur by chance 1 per cent of the time, if there were no association between religion and political allegiance. However, since our obtained value of χ^2 – 23.09 – is greater than 13.28, we conclude that the null hypothesis must be rejected in a two-tailed test. (In fact, our obtained value of χ^2 is significant even at the 0.1 per cent level, since it is greater than the corresponding critical value of 18.47.) On this evidence, there is almost certainly a statistical relationship between the two attributes in the population under study. If we choose, this conclusion can be followed up by performing one or more 'difference of proportions' tests (p. 110) on the extent of support for particular parties by those of each denomination. This would help to clarify where the main differences lie. However, it is noteworthy that two large contributions to the high value of χ^2 fall in the first row and this fact, taken together with the percentage distribution of Table 9.1, suggests that the significant result obtained derives partly from the fact that among Anglicans there is more extensive support for the Conservatives and correspondingly less support for Labour. A further notable feature is the relatively heavy support for the latter party among Roman Catholics.

The general procedure of the chi-square test follows this example closely. It is important to remember that the test is always performed on actual frequencies and never directly on percentages. An important restriction on the use of the chi-square test concerns the magnitude of the expected frequencies in the cells. This is, of course, dependent upon the total sample size (which preferably should be large) and the number of cells. The simplest case of all – that of the 2 x 2 table – is considered below, but in other instances (i.e. where there is more than one degree of freedom) the test may be used, provided that the expected frequency in at least 80 per cent of the cells equals or exceeds five and in the remaining cells exceeds one. When this condition is not met, perhaps because there are a large number of cells with small frequencies, it may be appropriate to proceed by combining some cells. However, this should only be done when a combination is meaningful. For instance, in our example both Baptists and Methodists had already been combined in the 'Free Church' category. However, it would make no sense in a study of political loyalties to combine (say) Liberals with those of no party allegiance. In this connection, one erroneous operation must always be avoided. This is the device whereby categories are combined in a contingency table simply so as to maximise the likelihood of obtaining a statistically significant result. This creates errors, because it increases the probability of rejecting valid null hypotheses and in general leads to a false statement of significance levels.

2 × 2 (or 'Fourfold') Contingency Tables

The simplest case of all occurs when there are just four cells in the contingency table* and it is then possible to make use of a straightforward formula. In this instance it is conventional to denote the observed frequencies by letters, as shown in Table 9.3. Then formula 9.1 becomes

$$\chi^2 = \frac{N(AD - BC)^2}{(A+B)(C+D)(A+C)(B+D)} \qquad (9.2)$$

with one degree of freedom. However, it is desirable to modify this formula by making what is referred to as a *correction for continuity*. This arises because the (theoretical) chi-square distribution is continuous, while the actual values which the computed χ^2 can take must be more limited, since the observed frequencies are necessarily whole numbers. When N is large, the errors which result from ignoring this fact are slight, but with smaller N (say under 40), it is essential to introduce a correction factor into the formula to provide a more accurate test of significance. The formula which should be used is

$$\chi^2 = \frac{N(|AD - BC| - \frac{1}{2}N)^2}{(A+B)(C+D)(A+C)(B+D)} \qquad (9.3)$$

where the vertical lines around $AD - BC$ mean (as usual) that the value of the expression is taken as positive regardless of whether it is calculated to

Table 9.3 *General 2 × 2 Contingency Table*

A	B	$A+B$
C	D	$C+D$
$A+C$	$B+D$	N

*As well as being handled as described here, this case may be treated using a difference of proportions test. In fact, the chi-square test is exactly equivalent to the two-tailed test for comparing two proportions.

be positive or negative. One can readily illustrate the ease with which the test may be performed in the 2 × 2 case.

Example. Suppose that in an investigation of the relationship between sex and political loyalties in a northern city the data shown in Table 9.4 were obtained from a random sample of 50 persons taken from the electoral roll. Our aim is to determine whether there is evidence of an association between sex and party support. Using formula 9.3, we find

$$\chi^2 = \frac{50(|70 - 270| - 25)^2}{22 \times 28 \times 25 \times 25}$$

$$= \frac{175^2}{7{,}700} = 3.98$$

In Table C in the Appendix with one degree of freedom, we see that a value in excess of 3.841 is required in a two-tailed test to reject at the 5 per cent level the null hypothesis, that there is no association between these two attributes. Since our calculated value is more than this, we do reject the hypothesis. None the less the nearness of the two values and the small sample size indicate that it would be wise to carry the investigation further by collecting additional data. In this connection it is worthwhile to know that χ^2 has a useful additive property. When several values have been computed from independent samples, in certain circumstances these may be summed to give a new chi-square value with degrees of freedom equal to the sum of the separate degrees of freedom. Suppose, for instance, we obtained data from three small independent random samples such as the above and all three showed a Conservative majority among women and a Labour majority among men. Then the three samples in aggregate might yield a significant result, even if only one or even none of the contingency tables did in isolation.

Table 9.4 *Sex and Political Party Support*

	Conservative	*Labour*	*Totals*
Men	7 (*A*)	15 (*B*)	22
Women	18 (*C*)	10 (*D*)	28
Totals	25	25	50

In the 2 × 2 case the use of the chi-square test is again subject to restrictions. When the total sample is greater than 40, it is appropriate to employ the test and the correction for continuity should be used, i.e. formula 9.3. When the sample size is between 20 and 40, the chi-square test may be used if all expected frequencies are 5 or more. In this latter case it is essential to use formula 9.3. On the other hand, with samples containing fewer than 20 members the chi-square test is inappropriate and its use would simply lead to error.

The Fisher Exact Test

Fortunately in the case of 2×2 tables when the sample size is small, a further test is available, which was introduced by R. A. Fisher. However, unlike the chi-square test, the Fisher test in its basic form is one-tailed and is used when the direction of a relationship has been predicted. With the cells labelled as in Table 9.3, the exact probability of the observed occurrence assuming the null hypothesis that there is no association between the attributes in the population under study is given by

$$P = \frac{(A+B)!(C+D)!(A+C)!(B+D)!}{N!A!B!C!D!} \tag{9.4}$$

The use of this formula can be illustrated by a simple example. Suppose that we are examining the relationship between trade union membership and support for the Labour Party in a defined population of manual workers. Our prediction is that membership and support will be positively associated and data obtained from a small random sample take the form displayed in Table 9.5. We seek to determine whether we can reject the null hypothesis of no association between the two attributes at the 5 per cent level. The exact probability of this occurrence assuming no association is determined by substituting the obtained values into formula 9.4:

$$P = \frac{12!7!9!10!}{19!9!3!0!7!} = 0.0024$$

This may look a complicated calculation but we can, of course, cancel certain factorials and simplify (remembering that $0! = 1$). However, with a modern calculator P can be obtained directly, without cancellation, by a single computation. In this instance, since P is less than 0.05 (and indeed even less than 0.01), our conclusion is that the null hypothesis of no association must be rejected.

The above example is comparatively simple, because one of the cells has

Table 9.5 *Trade Union Membership and Labour Party Support*

		Trade Union		
		Member	*Non-member*	
Labour Party	Support	9 (A)	3 (B)	12 ($A+B$)
	No support	0 (C)	7 (D)	7 ($C+D$)
		9 ($A+C$)	10 ($B+D$)	19 (N)

Table 9.6 *Trade Union Membership and Labour Party Support*

		Trade Union		
		Member	*Non-member*	
Labour Party	Support	8 (*A*)	4 (*B*)	12
	No support	1 (*C*)	6 (*D*)	7
		9	10	19

a frequency of 0. Should this not be the case, we must remember that more extreme deviations under the null hypothesis could occur with the same marginal totals. These more extreme cases need to be taken into account when we test the null hypothesis of statistical independence between the two attributes in the population sampled. As is usual with hypothesis testing we ask: What is the probability of the obtained distribution or one more extreme? For instance, suppose in the above example we had obtained instead the distribution shown in Table 9.6. Notice that this has the same marginal totals as Table 9.5 but the smallest obtained frequency is 1. If we seek to test the null hypothesis of no association on the data of Table 9.6, we must sum the probability of that particular occurrence with the probability of the more extreme one (shown in Table 9.5). This involves adding two probabilities, one of which has already been determined. The other, corresponding to the exact distribution of Table 9.6, is also given by formula 9.4:

$$P = \frac{12!7!9!10!}{19!8!4!1!6!} = 0.0375$$

Hence, the probability of the occurrence of Table 9.6 or an even more extreme one is

$$\text{probability} = 0.0375 + 0.0024 = 0.0399$$

and, given that this is less than the selected significance level of 0.05, we reject the null hypothesis and conclude that there is a positive association between trade union membership and support for the Labour Party among the manual workers constituting the population as a whole.

Measuring the Strength of Relationships

So far attention has focused on the issue of whether a relationship between two attributes exists. We have put forward the null hypothesis that there is no association and tested it. This leaves for consideration our second basic issue: determining the strength of a statistical relationship. This is a topic of prime importance in the application of statistics to the social sciences, for stronger relationships tend to be those having greater explanatory or predictive value. In this connection it may be believed at first sight that most of the work has already been done and that the significance level associated with a test may itself be taken as the substantive indicator: in other words, that if a test on one table yields statistical significance at the 1 per cent level, while that on another gives significance only at the 5 per cent level, then the former relationship is the stronger. However this is by no means necessarily the case, because the significance level achieved is dependent upon the sample size. In general when a relationship exists it is more easily *demonstrated* with larger samples than smaller ones.

Another, initially attractive, idea with contingency problems is that the value of the selected statistic – in particular, χ^2 – provides an indication of the degree of association. It is true that if the number of cases and the number of cells remain fixed, the larger the value of this statistic the greater is the degree of association. However, the usefulness of χ^2 as a measure is diminished by the fact that its value tends to be altered by changes in the number of cases or cells, even though the underlying relationship is fixed. It is especially noteworthy that from formula 9.1 it follows that if all observed frequencies are multiplied by some whole number r, then the new value of χ^2 is r times the former value. A consequence is that the chi-square statistic lacks any finite value for its upper limit, so that it is almost impossible to gain any intuitive grasp of what any particular value 'means'.

The Phi-Square Coefficient

Despite these problems, measures of association have been developed which are based on chi-square. Statisticians traditionally like to develop measures which vary either between 0 and 1, or between −1 and +1. The idea is that they should be intuitively meaningful and in particular that the value 0 should represent an absence of statistical association, while a value of unity should represent (in some sense) 'perfect association'. Since χ^2 varies directly with the number of cases N, the most straightforward measure of association based on it is standardised for N. Thus, the measure *phi-square* (ϕ^2) is defined by the formula:

$$\phi^2 = \frac{\chi^2}{N} \qquad (9.5)$$

In a general 2 × 2 contingency table labelled as in Table 9.3, it can be seen from formula 9.2 that ϕ^2 is given by

$$\phi^2 = \frac{(AD - BC)^2}{(A + B)(C + D)(A + C)(B + D)} \qquad (9.6)$$

In the 2 × 2 case ϕ^2 is a measure which is 0 when there is no relationship, and 1 when there is a 'perfect relationship'. The absence of a relationship implies that the proportional distribution of each row is the same and similarly for the columns. An example is provided by the contingency table

20 (A)	40 (B)	60
50 (C)	100 (D)	150
70	140	210

and it can be seen that ϕ^2 is 0, since the numerator of formula 9.6 becomes zero. On the other hand, the case of *perfect association* is one where a knowledge of either attribute is sufficient to predict the other. This is the case only when a pair of diagonally opposite cells contains zero values. The following table provides an example

100 (A)	0 (B)	100
0 (C)	50 (D)	50
100	50	150

and it can be confirmed that ϕ^2 is 1. Indeed, ϕ^2 only achieves its maximal value when there is perfect association. Given these properties, one may conclude that for the 2×2 contingency table ϕ^2 is a particularly useful measure.

Tschuprow's T

A problem, though, with ϕ^2 is that in the general case of a table with r rows and c columns the maximum value which it attains is greater than 1. Hence there is a need for somewhat more complex measures which, while still being based on χ^2, do have unity as their upper limits. The most popular is *Tschuprow's* T, defined by

$$T^2 = \frac{\chi^2}{N\sqrt{(r-1)(c-1)}} = \frac{\phi^2}{\sqrt{(r-1)(c-1)}} \tag{9.7}$$

Essentially this measure attempts to standardise chi-square with regard to both the number of cases N and the number of cells. The latter is achieved by the presence in the denominator of the number of degrees of freedom $(r-1)(c-1)$. Tschuprow's T only attains the upper limit of 1 when the number of rows equals the number of columns.

Coefficient of Contingency

A further measure of association developed on the basis of chi-square is Pearson's *contingency coefficient* C, defined by the formula

$$C = \sqrt{\frac{\chi^2}{\chi^2 + N}} \tag{9.8}$$

In this case the chi-square value in the numerator is being standardised both by N and the chi-square value in the denominator. The idea is that the number of cases can be standardised directly but the number of cells can be indirectly compensated for by the addition of χ^2. It is apparent that when there is statistical independence in a contingency table, since χ^2 is zero, C also is zero. However, a problem with this measure is that its upper limit is less than one. In fact, in a 2×2 table the maximum value of C is 0.707, in a 3×3 table it is 0.816, and in a 4×4 table, 0.866. In general the upper limit of C increases as the number of rows and columns increases. The conclusion must be that it is only wise to compare C-values resulting from contingency tables of the same size. Nevertheless this is an easily calculated and extremely useful measure.

The Chi-Square Test for 'Goodness of Fit'

We complete this chapter by describing a further important application of the chi-square test. Earlier, we examined how this test could be used to determine whether a set of observed frequencies differed sufficiently from a set of expected frequencies to reject the hypothesis under which the expected frequencies were obtained. Although at that stage attention focused on contingency tables, in fact the method is more general and can also be used to handle a range of problems involving single samples in which we seek to investigate whether the differences between observed and theoretically expected frequencies may be attributable to chance variations.

For example, as a result of sampling from a population, we frequently arrive at a frequency distribution which seems to approximate to the pattern of a known theoretical distribution, e.g. the normal curve. It would be very useful to know whether these empirical distributions are sufficiently close to the corresponding theoretical distribution that we may conclude that any observable differences are not significant but may simply arise from sampling. In the case of the normal distribution, it would have the particular advantage that we would then be justified in performing any other statistical tests on the data which depend upon the normality assumption. The method of testing for 'goodness of fit' is general and may be used, for instance, with the binomial and Poisson distributions, but we shall illustrate by reference to the normal curve because of its special importance.

Example. In a study of family decision-making a random sample of 210 families was selected within a city. The following data were gathered on the percentage of 'important' decisions of each family, which were judged to be jointly taken by husband and wife (over a specified period). Does the distribution significantly depart from what would be expected, if the corresponding population distribution were normal? Use the 5 per cent level of significance.

Percentage of decisions	*Frequency*
under 10.0	0
10.0–19.9	7
20.0–29.9	24
30.0–39.9	43
40.0–49.9	56
50.0–59.9	38
60.0–69.9	27
70.0–79.9	13
80.0–89.9	2
90.0 and over	0
	210

We need to determine the frequencies which we would expect to find in the various classes, if we actually had a normal distribution with the same arithmetic mean and standard deviation as our observed distribution. The initial step is to calculate these quantities from the sample data in the usual way, and it is found that $\bar{x} = 46.24$ and $s = 15.37$. We then proceed to determine the expected normal curve frequencies for each class. We do this by first determining z-values (i.e. standardised normal values; see p. 39) for the class boundaries. In this connection, it must be noted that the original percentages were determined to one decimal place (hence, values such as 9.96 would be rounded up to 10.0 and 19.94 rounded down to 19.9), so the class boundaries fall at 9.95, 19.95, 20.95 . . . 89.95. We calculate that 9.95 is $(9.95 - 46.24)/15.37 = -2.361$ standard deviations from the mean and, therefore, the corresponding z-value is -2.361. Similarly 19.95 corresponds to a z-value of $(19.95 - 46.24)/15.37 = -1.710$. The z-values of the various class boundaries appear as column 3 in Table 9.7.

Next from Table A in the Appendix, we construct column 4, which shows us the standardised normal curve area to the left of each z-value. To obtain the area *between* successive z-values (shown in column 5), we then calculate the *differences* of successive areas in column 4. At the top and bottom of column 5, we also record the tail areas to the left of the lowest z-value and to the right of the highest z-value, respectively. The final step is to multiply each area shown in column 5 by the sample size 210, and the expected normal curve frequencies are obtained. These are displayed in column 6 and can be compared by inspection with the observed frequencies shown in column 7.

In testing for goodness of fit, the last stage consists of determining χ^2 from formula 9.1, using the corresponding O and E values in columns 6 and 7 of Table 9.7. It should be noted that, in this case, we combine the first two classes and also the last two, for we must apply the rule that the chi-square test should not be used unless the expected frequency in at least 80 per cent of the cells equals or exceeds five, and in the remaining cells exceeds one. We have

Table 9.7 *Fitting a Normal Curve*

(1) *Classes*	(2) *Class boundaries*	(3) *z-values*	(4) *Cumulative normal areas*	(5) *Normal areas*	(6) *Expected normal curve frequencies*		(7) *Observed frequencies*	
under 10.0	9.95	−2.361	0.0091	0.0091	1.91	9.16	0	7
10.0–19.9	19.95	−1.710	0.0436	0.0345	7.25		7	
20.0–29.9	29.95	−1.060	0.1446	0.1010	21.21		24	
30.0–39.9	39.95	−0.409	0.3414	0.1968	41.33		43	
40.0–49.9	49.95	0.241	0.5952	0.2538	53.30		56	
50.0–59.9	59.95	0.892	0.8139	0.2187	45.93		38	
60.0–69.9	69.95	1.543	0.9386	0.1247	26.19		27	
70.0–79.9	79.95	2.193	0.9858	0.0472	9.91		13	
80.0–89.9	89.95	2.844	0.9977	0.0119	2.50	2.98	2	2
90.0 and over				0.0023	0.48		0	
				1.0000	210.01		210	

$$\chi^2 = \frac{(7-9.16)^2}{9.16} + \frac{(24-21.21)^2}{21.21} + \frac{(43-41.33)^2}{41.33} + \frac{(56-53.30)^2}{53.30}$$

$$+ \frac{(38-45.93)^2}{45.93} + \frac{(27-26.19)^2}{26.19} + \frac{(13-9.91)^2}{9.91} + \frac{(2-2.98)^2}{2.98}$$

$$= 3.76 \text{ (to two decimal places)}$$

As always with chi-square, we need to specify the number of degrees of freedom ν, and in fitting a normal curve as we have done this quantity is given by

$$\nu = k - 3$$

where k is the number of actual comparisons of frequencies (i.e. the number of items added to determine χ^2). In this case k is eight for although ten pairs of values were originally calculated in Table 9.7, the need to combine classes reduced the number of comparisons by two. It remains to refer to Table C in the Appendix, containing values for the χ^2 distribution and note that the critical value with 5 degrees of freedom is 11.07 at the 5 per cent level. Since the obtained value falls well short of this, we conclude that the normal curve provides a good fit and the original distribution could have arisen by random sampling from a normal population.

The justification for the formula for ν is similar to that given earlier in the case of the contingency table. In testing for normality, the expected values which we calculated were not all independent but had to satisfy three conditions. Their sum had to equal the sum of the observed frequencies, and the mean and standard deviation of the normal curve had to equal the mean and standard deviation of the observed distribution. Given that the number of conditions which were satisfied was three, the number of degrees of freedom ν is reduced by a corresponding amount. In general in a chi-square 'goodness of fit' test (e.g. on a binomial or Poisson distribution), the number of degrees of freedom equals the number of comparisons of observed and expected frequencies minus the number of quantities determined from the observed data which are used in the calculation of the expected frequencies.

Glossary

Chi-square (χ^2) test
Contingency table
Correction for continuity
Fisher exact test
The phi-square (ϕ^2) coefficient
Perfect association
Tschuprow's T
Coefficient of contingency (C)
'Goodness of fit' test

Exercises

1 The following table relates the religious denomination of a random sample of university students to their participation in non-religious

formal associations. Calculate the chi-square value and test for significance at the 5 per cent level:

	Participation			
	High	*Medium*	*Low*	*Totals*
Church of England	79	88	111	278
Free Churches	84	89	94	267
Roman Catholic	84	70	66	220
No religion	91	82	62	235
Totals	338	329	333	1,000

2 Two small random samples of the clientele in the private and public bars of a local public house were classified, on the basis of their occupations, into those with manual and those with non-manual employment:

	Non-manual	*Manual*	*Total*
Sample I (public bar)	1	6	7
Sample II (private bar)	6	2	8

It has been predicted that the public bar will be found to contain relatively more manual workers. Are these data compatible with the claim that manual and non-manual workers are to be found in equal proportions in the two bars?

3 A random sample of 71 adults was classified on the basis of sex and also according to the degree of political involvement. Is there any evidence from the following table to support the view that sex is statistically associated with political involvement? Also compute the value of the phi-square coefficient:

		Political Involvement		
		High	*Low*	*Total*
Sex	males	22	14	36
	females	14	21	35
		36	35	71

4 Test the following table derived from a random sample of 800 adults for association at the 1 per cent level and determine the coefficient of contingency C and Tschuprow's T:

		Style of Cooking Preferred				
		English	*French*	*German*	*Italian*	*Totals*
Nationality of subjects	English	67	16	79	67	229
	French	15	97	42	36	190
	German	16	32	133	49	230
	Italian	12	15	46	78	151
	Totals	110	160	300	230	800

5 Test to see whether the following distribution of attainment scores obtained from a random sample of 1,000 schoolchildren departs significantly from normality at the 1 per cent level:

Interval	*Frequency*
50.0–59.9	5
60.0–69.9	29
70.0–79.9	87
80.0–89.9	181
90.0–99.9	246
100.0–109.9	261
110.0–119.9	132
120.0–129.9	43
130.0–139.9	11
140.0–149.9	5
	1,000

Chapter 10

Regression and Correlation

Regression

Relationships between variables are very important in real-life decision-making. It is often vital to know, when the value of one variable is changed, how other variables are likely to be affected. For example, changes in child benefits can affect expenditure patterns, and possibly also the birth rate. Changes in tax levels on such things as tobacco, alcohol and petrol can directly or indirectly influence morbidity and mortality rates, as well as patterns of social life. The purpose of regression analysis, which is only carried out on intervally scaled variables, is to enable us to *predict* (in the sense of infer or estimate) the value of a particular variable, given the value(s) of one or more other variables. In this chapter we shall restrict ourselves to examining the case of two variable relationships. The two variables will be labelled x and y, the latter indicating the variable whose values are to be predicted. The actual analysis will always follow the same basic method, but it is essential for the social scientist to be aware of the different purposes of regression analysis and the various types of relationship which may be manifested in the data.

In some instances the main purpose of the analysis is to predict forward in time. This might be the case, for instance, where the attempt is made to relate the school performances of students (x) to their subsequent achievements in higher education (y), perhaps with the intention of improving student selection procedures. However, instead of the main purpose being predictive, it may be rather more explanatory as when measures of intelligence or motivation are related to attainment. A further point to consider is that there may or may not be a direct causal link between the variables under consideration. There is such a link between the level of incomes in the country (x) and expenditure on motor cars (y), but it is probably absent if one focuses instead on the level of incomes (x) and the number of convictions for driving offences (y). Again, there are instances where the link between two variables is to be accounted for by an antecedent causal influence or, alternatively, by some kind of 'feedback' or interactive process. An example of the latter would be the relationship between the extent of unemployment (x) and purchases of consumer durable goods (y), for it is an established fact that increased unemployment may reduce purchasing power, leading to a fall in the demand for consumer durables which in turn leads to further unemployment. In all these cases the y-variable may be predicted (with varying degrees of accuracy) for given values of the x-variable using the same method, but the underlying relationship may by quite different, and we need to be especially careful when we deal later with the strength of association between variables (correlation), that we do not make unjustified causal inferences. Generally, prediction is important but it is subordinate to the need to increase social scientific understanding.

One further point to note before we present the method, is that it is necessary to keep in mind the units of analysis. Sometimes the unit is a person, or an item. This would be so where we relate the educational attainments and incomes of individual people, or the demand and price for a particular product. Alternatively, the units may be groups or aggregates. In an example presented below the relation between the percentage of labour in agriculture (x) and the crude birth rate (y) is examined for a sample of twenty countries. The unit of analysis is important, because it affects the interpretation of any discernible regularities or calculated quantities.

Let us next consider what is involved in predicting values of a particular y-variable from those of an x-variable. Generally, the population is such that there is variation in the y-values associated with any given x-value (as is illustrated for sample data in Figure 10.1). In fact, what we have is a distribution of y-values for each x. In principle we could determine the mean value of y for each x, and this is what we would then use for predictive (or estimatory) purposes, i.e. $\bar{y}$ would be used as the predictor of y, given x. Now the curve (or path) traced out by $\bar{y}$ as x varies, which would be apparent in graphical portrayal, may be expressed by a mathematical equation, called the *regression equation of* y *on* x. Obviously, predicting on this basis is more reliable the smaller is the variation in y for a given value of x.

In this connection we generally have in mind indefinitely large populations, so for illustrative purposes it is easier to turn to samples. Suppose that we are engaged in a study of a sample of 1,000 manual workers employed by a large firm. For a particular year, we have obtained from the firm's records each worker's age (x) and the number of days he has been off work through illness (y). We can portray the data in graphical form by plotting the distribution of cases against the x- and y-axes of rectangular co-ordinates, and by this means we obtain a *scatter diagram*. Given that age is strictly a continuous variable, with sample data such as this no two ages (calculated to a high degree of accuracy) would recur, so there would be no variation in y for a particular x. However, what we have done in Figure 10.1 is to group ages in five-year intervals so that many cases occur in each interval and then portray the associated y-distributions. To simplify the diagram even further, we have taken each dot in the figure to represent five workers.

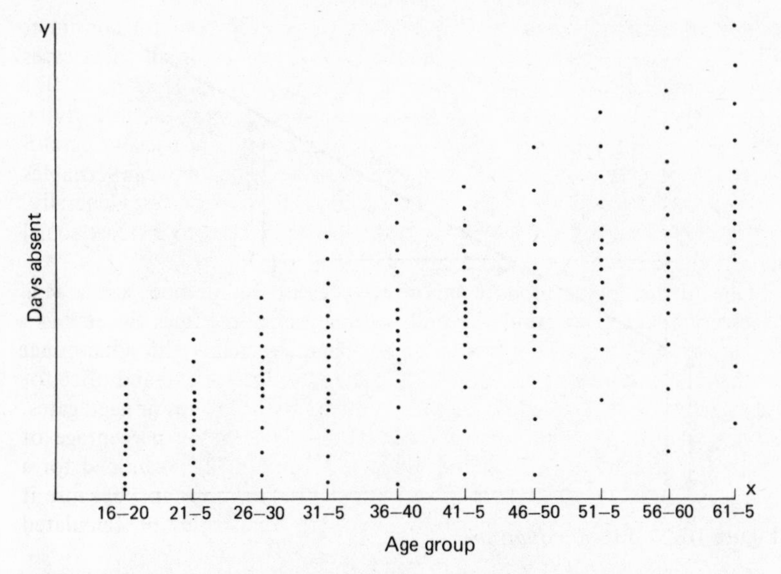

Figure 10.1 *Scatter diagram of age (x) against days absent (y).*

It is important to scrutinise the scatter diagram carefully, as it can very often suggest the type of relationship and hence the form of the regression equation. In this case, we can see immediately that there is a tendency for more days to be lost as age increases, i.e. y tends to increase with x. Also, although y varies for each group of x-values, we can also discern some clustering perhaps suggesting the shape of a line or curve. If we choose, we can determine the variance in y for each value of x and then sum this quantity over the x-values. The greater this sum as a proportion of the total variance of all y-values, the less useful is the x-variable as a predictor of y. If, however, the association between x and y is strong with the data points all clustering about a particular curve, then for any value of x, we may accurately predict what the corresponding y-value is without having to measure it. This may also assist in the interpretation of underlying causal relationships.

Determining the regression equation for a population directly in the way implied by the earlier definition, is often very difficult or impossible. Instead, we have to proceed indirectly by first postulating that it takes a certain form and in this way narrowing down our search. This is very much in line with other studies of the workings of our society and economy, where theoretical models are suggested and then tests are performed to determine how well these models fit real-life data. Some models may be accepted as good reflections of a social process, others will be rejected as inadequate. Here we have already limited the discussion to two variables, and we shall now restrict it even further by considering cases where the population regression equation corresponds to a straight line, i.e. the

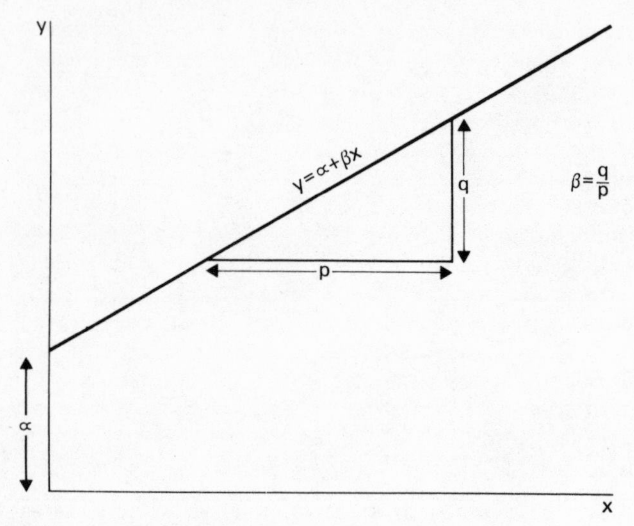

Figure 10.2 *Linear equation.*

equation is of the form

$$y = \alpha + \beta x \tag{10.1}$$

In this case our model is termed linear.* α is termed the *intercept* which the line makes with the y-axis, i.e. the value of y when $x = 0$; β is the *slope* of the line, i.e. the increase in y for a unit increase in x (Figure 10.2). In the methods which we go on to describe for handling regression problems and estimating α and β, we also need to assume that the distributions of the y-values for each x are normal, and that the variances of the y-distributions are the same for each value of x.

Since we are extremely unlikely to encounter variables for which all the sample data points actually lie on a straight line, the general approach that is adopted is to obtain the equation of the line which 'best fits' the data. We could do this by eye, using a ruler and drawing a line so that it follows the general direction of the points and divides them so that approximately half lie above the line and half below. This method, although perhaps familiar from elementary science classes is not at all reliable in that two people might draw the line in different positions. What we need is a criterion by which, with the same data, everyone will get exactly the same line of 'best fit'. The most commonly used and the most theoretically sound is the *least-squares criterion*. To explain this, let us refer to Figure 10.3. Each point is represented in the (x,y) form of rectangular coordinates. The equation of the line we are seeking as an approximation to that of the population regression line given in expression 10.1 is

* Below we also consider a case where a non-linear model can be so transformed (or modified) as to exhibit a linear form.

$$y = a + bx \tag{10.2}$$

where a and b are initially unknown numerical values, which when determined will fix the position of the line. These values of a and b will themselves only be estimates of the population values α and β, and if we took another set of sample values for our two variables, we should undoubtedly obtain somewhat different numerical values for a and b. This is analogous to taking two random samples from a population with mean μ and variance σ^2, where both pairs of statistics $\bar{x}_1, s_1^2$ and $\bar{x}_2, s_2^2$ are estimates of μ and σ^2.

What we focus on in connection with the regression of y on x are the deviations of the data points from the hypothetical line of best fit, and our interest centres on the deviation in the vertical (y) direction. Figure 10.3 shows vertical lines constructed from the data points to meet the best-fitting line. The points where they meet it are subject to the equation $y = a + bx$. So if the data points have co-ordinates (x_1, y_1), (x_2, y_2) . . . (x_n, y_n), and the points on the line are designated by (x_1, Y_1), (x_2, Y_2) . . . (x_n, Y_n), then the Y-values are given by $Y_1 = a + bx_1$, $Y_2 = a + bx_2$, and so on. Hence, the vertical deviation from the point (x_1, y_1) to the line will be the difference between y_1 and Y_1, i.e. $y_1 - (a + bx_1)$. Corresponding expressions apply for the other deviations, e.g. $y_2 - (a + bx_2)$.

Since some points are above and others below the line, some deviations will be positive and others negative. In seeking to obtain a line of best fit, it is natural to attempt to minimise the deviations as a whole. It would not

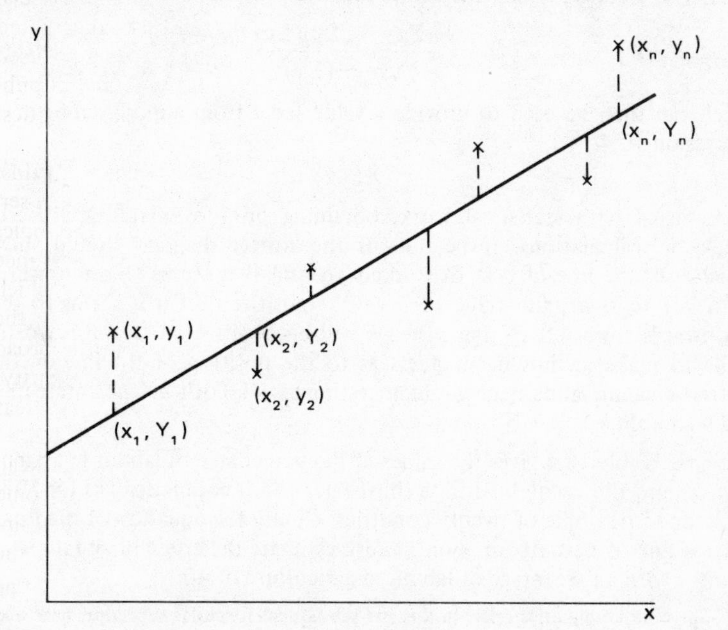

Figure 10.3 *Deviations from line of best fit.*

do however to minimise their sum, since positive and negative values might cancel out to produce a small overall value (or even zero) for a poorly fitting line. The least-squares criterion gets round this problem, because what one aims to minimise instead is the sum of squares of deviations, namely,

$$\Sigma(y_i - a - bx_i)^2 \tag{10.3}$$

It can be shown that this criterion defines a unique line, which has the property that not only is the sum of the deviations zero, but also the corresponding standard deviation is a minimum.

So we need to determine a and b such that expression 10.3 is a minimum. This can readily be done, using differential calculus, and one arrives at a pair of simultaneous equations in a and b which may then be solved. They are known as *normal equations:*

$$\Sigma y_i = an + b\Sigma x_i \tag{10.4}$$

$$\Sigma x_i y_i = a\Sigma x_i + b\Sigma x_i^2 \tag{10.5}$$

The quantities Σy_i, Σx_i and Σx_i^2 are obtained from the sample data by adding, respectively, the y-values, the x-values and the squares of the x-values. $\Sigma x_i y_i$ is obtained by multiplying together corresponding x- and y-values and adding the resulting products; n is of course the number of pairs of observations. We can solve the equations by multiplying expression 10.4 by Σx_i and expression 10.5 by n, then subtracting one from the other. We, thereby, obtain the computationally straightforward formula*:

$$b = \frac{n\Sigma xy - (\Sigma x)(\Sigma y)}{n\Sigma x^2 - (\Sigma x)^2} \tag{10.6}$$

which can then be used to provide a value for a from a modified form of expression 10.4:

$$a = \bar{y} - b\bar{x} \tag{10.7}$$

As usual with statistical work, obtaining prior estimates is a useful check on calculations. Inspection of the scatter diagram should show which way the line of best fit is likely to run. If it seems to run upwards from left to right, the value of b will be positive, but if it seems to run downwards from left to right, then b will be negative. In fact, it is worthwhile to make an intelligent guess as to the position of the line on the scatter diagram and, hence, obtain estimates of both a (the intercept), and b (the slope).

Example. Table 10.1 gives the values of the percentage of labour in agriculture, x, and the crude birth rate (births per 1,000 population, y) for 1970 for a random sample of twenty countries. Obtain the equation of the least-squares line of best fit for y on x. Also estimate the crude birth rate for a country with a percentage of labour in agriculture of 40.

*Since no ambiguities arise, in this and subsequent formulae subscripts have been dropped.

Table 10.1 *Data for Twenty Countries on Agricultural Labour and the Birth Rate*

Country	*Percentage labour in agriculture* x	*Crude birth rate* y	*Country*	*Percentage labour in agriculture* x	*Crude birth rate* y
South Korea	50	31	Australia	8	20.5
Malaysia	55	37	Chile	24	28
Pakistan	68	51	Turkey	62	40
France	15	16.7	Norway	18	16.6
Italy	21	16.8	Zambia	81	50
Spain	34	19.6	Ghana	56	47
UK	3	16.2	Tunisia	41	42
Brazil	52	38	Kenya	88	48
Canada	6	17.5	Japan	17	19
Iran	42	41	USSR	31	17.4

We notice immediately that not all y-values are provided to the first decimal place; generally speaking developed countries provide more accurate figures. This is an unsatisfactory feature of the source material, but the resultant (small) error is unavoidable. We have $n = 20$, and we obtain the following sums:

$$\Sigma x = 772 \qquad \Sigma x^2 = 41{,}664 \qquad \Sigma y = 613.3 \qquad \Sigma xy = 29{,}279.5$$

From expression 10.6,

$$b = \frac{20 \times 29{,}279.5 - 772 \times 613.3}{20 \times 41{,}664 - (772)^2} = 0.4725$$

and from expression 10.7,

$$a = \frac{613.3}{20} - 0.4725 \times \frac{772}{20} = 12.43$$

The equation of the line we are seeking is therefore

$$y = 12.4 + 0.47x$$

On the scatter diagram (Figure 10.4) this line has been drawn in and also an estimate of its position made before the calculations were performed.

Turning next to the problem of estimating a specific y-value, we substitute an x-value of 40 into our equation (with values for a and b recorded with appropriate accuracy), and find:

$$y = 12.43 + 0.4725 \times 40 = 31.3$$

Clearly, our estimate is only accurate to the nearest whole number (given the input data), so the required estimate of the crude birth rate is 31.

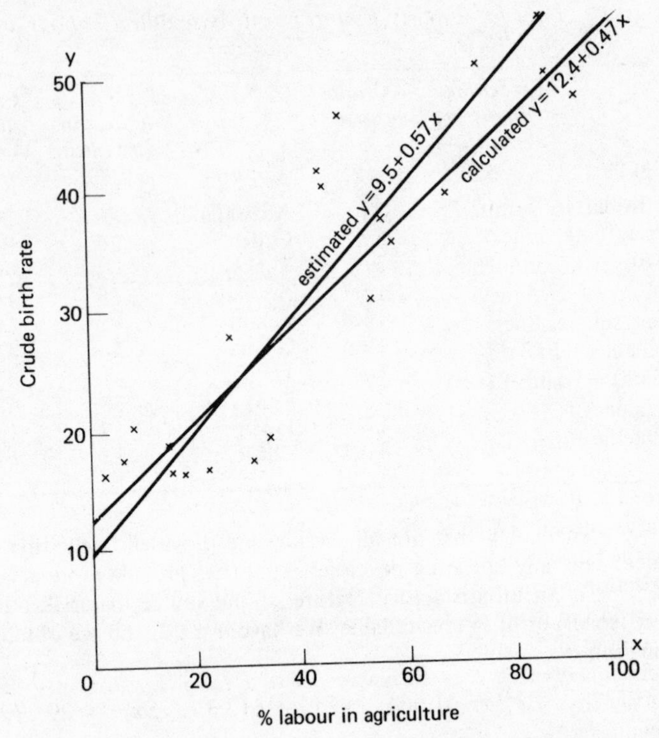

Figure 10.4 *Relation between birth rate and percentage employed in agriculture.*

Correlation

After fitting a least-squares line, it is important to inquire how *well* it fits the data. This is an expression of the policy that one must not simply present models, but also evaluate them. There is also the vital point to consider that the above procedure for obtaining lines of best fit guarantees that a line will be obtained as long as there is variation in x and y, despite the fact that the 'fit' may be poor. It would clearly be nonsense to base arguments concerning the underlying relationship on the outcome, if the model is manifestly an inappropriate one.

To assess how well the line fits, we need a new measure of association called the *Pearson product-moment correlation coefficient*, which is given the symbol ρ (rho) when referring to a population and r for sample data. The basic defining formula, given n pairs of x- and y-values, is

$$r = \frac{\Sigma(x-\bar{x})(y-\bar{y})}{\sqrt{[\Sigma(x-\bar{x})^2][\Sigma(y-\bar{y})^2]}} \tag{10.8}$$

and we also have the equivalent forms

$$r = \frac{[\Sigma(x-\bar{x})(y-\bar{y})]/n}{\sqrt{[\Sigma(x-\bar{x})^2/n][\Sigma(y-\bar{y})^2/n]}} \tag{10.9}$$

and

$$r = \frac{n\Sigma xy - (\Sigma x)(\Sigma y)}{\sqrt{[n\Sigma x^2 - (\Sigma x)^2][n\Sigma y^2 - (\Sigma y)^2]}} \tag{10.10}$$

with the latter formula being computationally the most straightforward with raw data. Indeed, inspection of formula 10.10 and consideration of equations 10.6 and 10.7 shows that all the relevant sums with the single exception of Σy^2 will have been obtained in calculating the equation of the least-squares line.

Pearson's r has the property that it can take any value between (and including) -1 and $+1$. Its sign will be that of the slope of the least-squares line. Values of -1 and $+1$ imply that all the data points actually lie on the line. On the other hand, if two variables are independent and the dots on the scatter diagram fall randomly, then the value of r will approximate to zero, i.e. it will only deviate from zero because of chance factors. In summary, one can say that the better the fit to the least-squares line, the greater the (absolute) magnitude of r.

Inspection of formula 10.9, shows that in that case under the square root sign in the denominator, we have the variance of x multiplied by the variance of y. In the numerator of the formula is an expression of a similar form which is referred to as the *covariance* of x and y. So, in words, we are able to say that r is given by the ratio of the covariance to the product of the (unadjusted) standard deviations of x and y.

Example. Calculate r for the data presented earlier on percentage of labour in agriculture and the crude birth rate for twenty countries. We shall use the computational formula 10.10. All relevant sums were obtained earlier except $\Sigma y^2 = 22{,}090.55$. So:

$$r = \frac{20 \times 29{,}279.5 - 772 \times 613.3}{\sqrt{[20 \times 41{,}664 - (772)^2][20 \times 22{,}090.55 - (613.3)^2]}}$$

$$= 0.898$$

Significance testing. It is important to realise that when different random samples from a population are taken, there will be variation in the associated values of r. Though r may be used as an estimator of ρ, particular r-values may diverge considerably from ρ. In fact, even when the x- and y-variables are independent in the population and hence $\rho = 0$, with quite small samples it is possible to obtain r-values of 0.5 or greater. To determine whether or not we have good reason to believe ρ is non-zero, involves specifying the sampling distribution of r.

To put the procedure formally, we seek to test the null hypothesis H_0 that $\rho = 0$ against an alternative H_1. The latter may be non-directional and of the form $\rho \neq 0$, or directional so that either $\rho < 0$ or, alternatively, $\rho > 0$. Under H_0 the sampling distribution of r is as one would expect – symmetrical about zero. When n is large (say over 50), it is closely approximated by a normal distribution with mean zero and variance $1/(n-1)$, so under the null hypothesis

$$\frac{r}{1/\sqrt{n-1}} \tag{10.11}$$

is approximately the standardised normal variable z. On the other hand, when n is smaller, the sampling distribution of r is by no means normal, but in these circumstances we find that the statistic

$$r\sqrt{\frac{n-2}{1-r^2}} \tag{10.12}$$

is distributed as Student's t with $n-2$ degrees of freedom. So in a given case, we specify the significance level α, calculate a value for the appropriate statistic and determine whether or not it falls in the rejection region.

To illustrate, let us return to our data from a random sample of 20 countries and test at the 1 per cent level $H_0: \rho = 0$ against the non-directional alternative $H_1: \rho \neq 0$. We calculated r to be 0.898. This is a small-sample problem, so substituting into formula 10.12, we obtain

$$0.898\sqrt{\frac{20-2}{1-0.898^2}} = 8.66$$

We observe from Table B in the Appendix that in a two-tailed test with $\nu = 18$, the rejection region in the positive direction lies beyond the critical t-value of 2.878. Since this has been exceeded, we reject H_0 and conclude that there is a positive association between percentage labour in agriculture and the crude birth rate.

Above we have seen how to obtain values for a, b and r and also how to perform a significance test for r. However, as with other cases where we obtain estimates from sample data, it is possible to construct confidence intervals. The methods are beyond the scope of this book and for an account see Blalock, 1972, pp. 400–405; Yeomans, 1968, vol. 2, pp. 219–22.

Non-Linear Models

Although our account has focused on linear models, it is worth briefly noting that there are a number of non-linear models for which the calculations are essentially the same, once one or both variables have been modified or transformed by some simple mathematical operation. One of the more frequently occurring in the social sciences is called the *exponential model*, which has applications to many growth situations evident in the population and economic fields. The model is of the form:

$$y = AB^x \tag{10.13}$$

Usually x is the time variable, and what one seeks to do is to use the least-squares criterion in the analysis of trends.

In the case of formula 10.13 the method is rather easier than might be anticipated. Taking logarithms on both sides of the equation, we obtain:

$$\log y = \log A + x \log B$$

Putting $Y = \log y$, $a = \log A$ and $b = \log B$, this becomes $Y = a + bx$. We now have a linear model for which we can determine values for a and b via the normal equations 10.4 and 10.5, namely

$$\Sigma Y = an + b\Sigma x$$

$$\Sigma xY = a\Sigma x + b\Sigma x^2$$

Having solved for a and b in the usual way, we then take antilogarithms to arrive at values for A and B.

Example. The population of Wales for the Census years from 1831 to 1921 is given below. Taking the x-variable as the number of years after 1831, and the y-variable as the population in thousands obtain a least-squares curve of best fit for the data:

Year	1831	1841	1851	1861	1871	1881	1891	1901	1911	1921
x	0	10	20	30	40	50	60	70	80	90
Population (thousands) y	904	1,046	1,163	1,286	1,413	1,572	1,776	2,013	2,421	2,656

Without reference to the nature of the variables, we might be forgiven for taking the linear model $y = a + bx$ and fitting that, for inspection of the data and its graphical representation in Figure 10.5 suggest that the fit would be quite good. If we carry out the necessary calculations, the result (which the reader may care to confirm) is obtained that the equation of the least-squares trend line is

$$y = 772.1 + 18.95x$$

and the correlation coefficient r is 0.978, a very high value.

However, if we reason that the model for population trends is more likely to be an exponential growth model (since the magnitude of births, deaths and to a lesser extent migration tends to depend upon overall population size), then we may fit instead $y = AB^x$ to the data. We start by determining $Y = \log y$ values (taking logarithms to the base 10):

x	0	10	20	30	40	50	60	70	80	90
Y	2.956	3.020	3.066	3.109	3.150	3.196	3.249	3.304	3.384	3.424

To these values, we fit $Y = a + bx$, and we obtain by the usual method

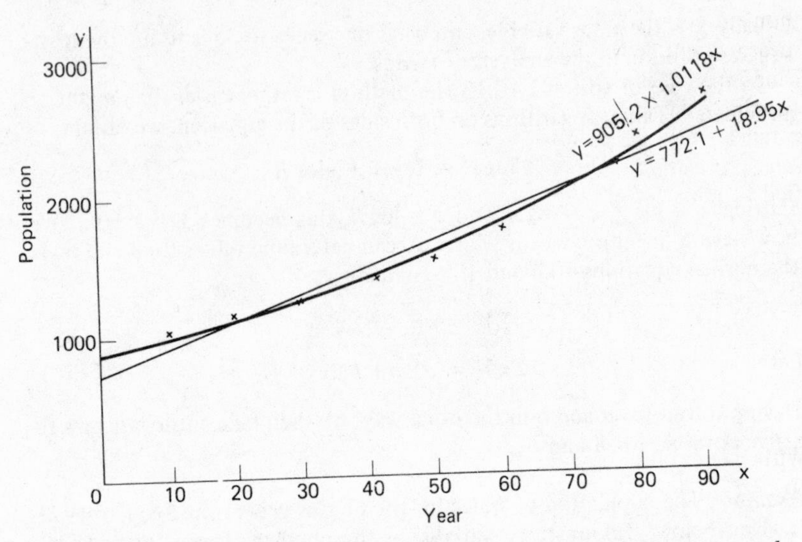

Figure 10.5 *Use of linear and exponential models for population growth.*

$a = 2.95627$, $b = 0.005101$ with an associated $r = 0.997$. Now $a = \log A$ and $b = \log B$, so taking antilogarithms, we obtain $A = 904.2; B = 1.0118$, and the required equation is

$$y = 904.2 \times 1.0118^x$$

The value of B indicates that there was an approximate rate of growth of population of 1.2 per cent per annum over the period. We note that at 0.997 the correlation coefficient was even higher than the value obtained for the linear model. This illustrates the need to examine the variables carefully before deciding on the most appropriate model. It is the fact that the growth rate is fairly low that has the consequence that the straight line and exponential models both fit the data very well.

Because it is now quite easy for social scientists to gain access to computers and associated statistical 'packages', there is a temptation when looking for a suitable model to fit to one's data, to try out a number and select as best that one which gives the highest correlation coefficient. The problem then becomes one of explaining in social scientific terms why that particular model should fit the variables in question. This may well be extremely difficult and the alternative approach of selecting at the outset a model which is interpretable – like the exponential growth model in our example – seems preferable.

Rank Correlation

As we have seen, various conditions need to be met if one is to carry out a least-squares linear regression analysis and obtain the Pearson product moment correlation coefficient, including the requirement of interval scale measurement. When instead we have ordinally scaled data, although we cannot proceed in this way, we are still able to analyse correlations. Essentially what we do when we have n pairs of ranks is to substitute the x ranks and y ranks into formula 10.8, and we then obtain the *Spearman rank correlation coefficient* r_s. However, in practice we might use a simplified formula for r_s. With the xs and ys as ranks and putting $X = x - \bar{x}$ and $Y = y - \bar{y}$, we obtain from formula 10.8

$$r_s = \frac{\Sigma XY}{\sqrt{\Sigma X^2 \Sigma Y^2}}$$

Let us now put $d = x - y$, and since with ranks $\bar{x} = \bar{y}$, we also have

$$d = (X - \bar{x}) - (Y - \bar{y}) = X - Y .$$

With this notation the above formula for r_s can be shown to be equivalent to

$$r_s = \frac{\Sigma X^2 + \Sigma Y^2 - \Sigma d^2}{2\sqrt{\Sigma X^2 \Sigma Y^2}} \qquad (10.14)$$

If we then express ΣX^2 and ΣY^2 in terms of n, this reduces to the computationally useful

$$r_s = 1 - \frac{6\Sigma d^2}{n^3 - n} \qquad (10.15)$$

Table 10.2 *Data on Popularity and Ability of Pupils for Determination of Rank Correlation*

Pupil	*Popularity rank* x	*Ability rank* y	$d = x - y$	d^2
A	3	1	2	4
B	9	12	−3	9
C	1	2	−1	1
D	11	7	4	16
E	6	10	−4	16
F	14	15	−1	1
G	2	3	−1	1
H	5	8	−3	9
I	13	9	4	16
J	10	11	−1	1
K	4	6	−2	4
L	7	4	3	9
M	8	5	3	9
N	12	13	−1	1
O	15	14	1	1
			$\Sigma d = 0$	$\Sigma d^2 = 98$

Example. Fifteen school pupils were selected randomly and ranked by a sociologist in terms of their popularity in class (rank 1 indicating the greatest popularity) and also according to their academic ability as determined by the teacher (rank 1 indicating the greatest ability) (Table 10.2). To what extent are these two aspects correlated?

From formula 10.15, given that $n = 15$, we have

$$r_s = 1 - \frac{6 \times 98}{15^3 - 15} = 0.825$$

A useful computational check is that Σd must always equal zero.

Ties. Sometimes individuals or items cannot be distinguished on a given scale. When this occurs, they are each given the mean of the ranks they would have received had there been no ties. For instance, if there is a three-fold tie after the fifth rank, then each item receives a rank of $(6 + 7 + 8)/3 = 7$. With this adjustment, if there are only a few ties, one can continue to use formula 10.15. However, if there are many ties, one should use instead a modified form of formula 10.14.

To do this, one first takes the x ranks and determines for each group of ties the quantity $T_x = (t^3 - t)/12$, where t is the number of items tied for a given rank. One then sums for all tied groups to obtain ΣT_x. In the same way ΣT_y is obtained by considering groups of tied y ranks. One next determines the two quantities

$$\Sigma X^2 = \frac{n^3 - n}{12} - \Sigma T_x \quad \text{and} \quad \Sigma Y^2 = \frac{n^3 - n}{12} - \Sigma T_y$$

which are finally substituted, along with the calculated value of Σd^2, directly into formula 10.14.

Significance testing. To test the null hypothesis of no correlation in the population we can proceed, when n is 15 or larger, in the same way as we did for the test of the product moment correlation coefficient r. However, none of the assumptions underlying least-squares linear regression and correlation need apply. What we do is to determine the quantity

$$r_s\sqrt{\frac{n-2}{1-r_s^2}} \qquad (10.16)$$

If the null hypothesis is true, then this statistic is approximately distributed as Student's t with $n - 2$ degrees of freedom. In the case of the above example formula 10.16 gives

$$0.825\sqrt{\frac{13}{1-0.825^2}} = 5.26$$

Suppose we have selected the 0.1 per cent significance level. Since the calculated value exceeds 4.221, which is the critical t-value with $\nu = 13$ in a two-tailed test at the 0.1 per cent level (see Table B in the Appendix), we conclude that there is very good reason to believe that a positive association exists in the population between popularity and ability.

Case Study in Regression and Correlation

In 1968 notes were being prepared for a first-year university statistics class on regression and correlation. Data on the number of casualties in road accidents and the number of currently licensed vehicles in Wales were taken from the *Digest of Welsh Statistics*. The two variables seemed suitable for the purpose of illustrating regression and correlation techniques, particularly as there seemed to be good reason to expect that an increase in the number of vehicles, resulting in increased congestion on the roads, would serve to influence the number of casualties. Values for both variables were taken for the years 1958–67 inclusive, and these appear in Table 10.3. The scatter diagram for the data is shown in Figure 10.6.

One can see that the general direction of the data points is upwards as x increases following a reasonably linear path. However, we note that casualties fell a little from 1960 to 1962 and from 1965 to 1967. The latter period is also notable for the fact that it includes the only instance where a pause in the increase in vehicle numbers was experienced – from 1965 to 1966. Nevertheless the broad picture is clear and calculation yields the following totals:

$$\Sigma x = 5{,}355 \qquad \Sigma y = 163.7 \qquad \Sigma xy = 89{,}261.4 \qquad \Sigma x^2 = 2{,}962{,}623$$

The equation of the least-squares line obtained from equations 10.6 and 10.7 is

$$y = 7.35 + 0.0168x$$

with a correlation coefficient r of 0.94. Drawing in the regression line on

Table 10.3 *Welsh Data on Vehicles and Road Casualties, 1958–67*

Year	*Number of vehicles* x *(thousands)*	*Number of casualties* y *(thousands)*
1958	385	13.5
1959	415	14.7
1960	451	15.6
1961	477	15.4
1962	511	15.3
1963	563	15.8
1964	596	17.8
1965	638	18.8
1966	637	18.5
1967	682	18.3

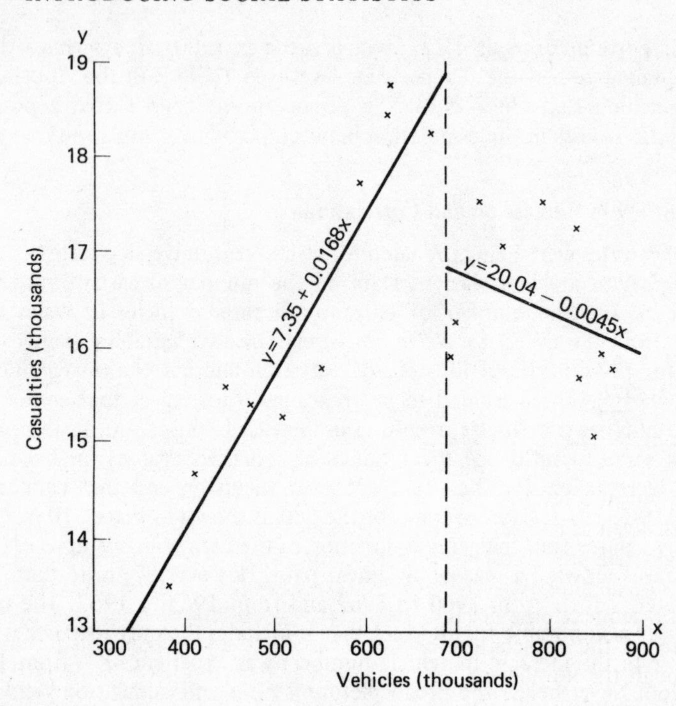

Figure 10.6 *Least-squares lines of best fit.*

the scatter diagram (Figure 10.6), it is quite evident that there is a high correlation as none of the data points is far from the line. So far, so good. The example was incorporated into the statistics notes.

At the end of 1967 an important policy change occurred which was to have a profound effect on casualties: the breathalyser was used to detect drunken drivers. This seemed to bring about a marked change in drivers' (bad) habits, because although the number of vehicles increased to 695 thousand in 1968, casualties fell to 15.9 thousand (see Table 10.4). Now, in regression studies of this kind there is a need to be aware that the relevance of particular models depends upon the two variables continuing to be causally related in the same kind of way as they have in the past. Causally significant but unpredictable events can have a major effect on the appropriateness of models and the adequacy of predictions about the future. There is a need to reappraise the situation and attempt to determine what the future significance of such events is likely to be.

In this case, given the casualty figures for 1969 and 1970, it seemed possible that the effect of the breathalyser might have been to pull the regression line downwards, i.e. reduce the value of the constant a in the regression equation by about 3, but leave the slope b unaltered. This would tie in with the expectation that the casualty figures would keep on

Table 10.4 *Welsh Data on Vehicles and Road Casualties, 1968–77*

Year	*Number of vehicles* x *(thousands)*	*Number of casualties* y *(thousands)*
1968	695	15.9
1969	699	16.3
1970	729	17.6
1971	754	17.1
1972	796	17.6
1973	836	17.3
1974	868	16.0
1975	860	15.1
1976	878	15.8
1977	840	15.7

rising but at a level some 3,000 below the values predicted by the original model. Even as late as 1973, it still seemed possible that this would happen as slight falls in the casualty figures from 1970 to 1971 and 1972 to 1973 were not dissimilar to those observed previously. However, once the 1974 and subsequent data became available, it was quite obvious that the very nature of the association between casualties and vehicles had changed. Although the number of vehicles increased over the 1968–77 period from 695 thousand to 840 thousand, casualties actually fell from 15.9 to 15.7 thousand. The equation of the regression line for this period is

$$y = 20.04 - 0.0045x$$

with a correlation coefficient r of −0.36 (Figure 10.6). Hence, the apparent simplicity of this exercise in linear regression had been disrupted. Quite properly there was a need to rewrite the statistics notes to reflect the new-found complexity!

What had happened in those years to disrupt the model so much? It seems unlikely that the breathalyser could have been itself a sufficient cause. It has been suggested that the more widespread use of car seat-belts after 1973 might have played its part, together with lower mileages consequent upon the substantial petrol price rise. Doubtless future data will shed some light on the problem and certainly they can help us to develop a more complex and appropriate model.

Glossary

Regression equation
Scatter diagram
Line of best fit
Least-squares criterion
Normal equations
Pearson product-moment correlation coefficient
Spearman rank correlation coefficient

Exercises

1 An investigation was carried out into the time spent by final-year undergraduate students on non-academic activities and its relationship with their academic performance. As part of the investigation twenty students chosen at random were asked to determine how many hours per week on average during term they spent on non-academic activities. These figures together with their subsequent examination marks were as follows:

time spent (hours)	10	5	12	9	10	14	4	6	11	9
examination marks (%)	53	61	48	46	60	51	68	70	41	49
time spent (hours)	16	8	10	6	12	8	9	4	13	9
examination marks (%)	38	62	57	64	45	54	54	65	42	56

Draw a scatter diagram. Obtain the equation of the least-squares line of best fit of marks against time spent, and calculate the value of the product-moment correlation coefficient. Test whether the latter value is significantly different from zero at the 5 per cent level. Comment on your results.

2 The data below concern the percentage of pregnant women attending clinics for antenatal examinations (x) and the perinatal mortality rate (stillbirths and deaths of infants under one week per 1,000 births, y) for Wales, 1966–73:

	1966	1967	1968	1969	1970	1971	1972	1973
x	44.7	41.1	35.4	30.1	28.3	30.4	29.1	25.5
y	30.1	27.8	27.5	26.7	25.5	24.4	22.3	21.4

Obtain the equation of the least-squares line of y on x and also the product-moment correlation coefficient. Attempt to interpret your results.

3 The numbers of completions of houses in the public sector (x) and the private sector (y) in Wales for the years 1960–74 are shown below. Calculate the value of the product-moment correlation coefficient and comment on your result:

	x	y
1960	5,526	6,078
1961	5,650	7,019
1962	7,609	7,501
1963	6,471	7,609
1964	9,207	9,762
1965	10,023	9,501
1966	9,743	9,617
1967	10,936	9,222
1968	9,233	9,949
1969	7,998	9,306
1970	6,825	8,648
1971	5,927	9,174
1972	4,135	10,635
1973	3,377	10,957
1974	3,674	8,137

4 A random sample of fifteen regulars at a public house was selected by a sociologist. After interviewing them, the sociologist ranked the members of the sample according to their standing in the local community (rank 1 indicating the highest status). Subsequently he also ranked them according to their average weekly alcohol consumption (rank 1 indicating the highest consumption). In each case when individuals tied for a rank, they were given the mean of what would have been the associated untied ranks. From the following data, determine the rank correlation coefficient and test to see whether its value is significantly different from zero at the 0.1 per cent level, given that it has been predicted that the correlation will be negative:

Sample member	A	B	C	D	E	F	G	H	I	J	K	L	M	N	O
Status rank	4	1.5	8.5	4	1.5	6	11.5	8.5	14	8.5	14	4	14	11.5	8.5
Alcohol rank	15	12	9	14	13	10	7.5	7.5	1	3	2	11	4	6	5

Chapter 11

Sources of Statistics

Throughout this book methods of analysing data have been presented. In this chapter, we examine certain sources of data to which these methods may be applied. Two options are available to the social investigator: he can collect and assemble data himself, or he can use data made available by others. If he adopts the former course of action, he may need to survey the whole or part of a defined population under study. This will involve designing a questionnaire or interview schedule, specifying precisely the population to be covered, choosing a sampling procedure (should he decide to contact only part of the population) and organising the process of data collection. (For a detailed account of survey methods, see Moser and Kalton, 1971.) The main alternative to this potentially expensive and time-consuming operation is to gain direct access to already-existing data of the required type. Such data could conceivably be made available privately or it could be derived from a data bank, for instance, that containing social survey material sponsored by the Social Science Research Council, situated at the University of Essex.* However, perhaps the most frequently used source of readily available data relating to social affairs is government publications, and it is the character and scope of these 'official statistics' which warrant particular attention here.

The extensive administrative needs of government embrace many fields, of particular importance being industry and economic planning, social service provision and the administration of housing and education. The topics covered in official statistics reflect a governmental view of policy relevance or the public interest, and the quantitative data generated are those which are generally required for, or yielded by, administrative processes. In the social field there are extensive series of published statistics in the broad fields of population (including vital statistics), health, social security, housing, education and crime, as well as some coverage of narrower areas such as accidents, the fire service and tourism.

A notable feature of many official statistics is that they provide head counts. This meets the needs of government departments and is in some respects suited to the requirements of those involved in developing or implementing social policy. However, this feature is a limitation from the

*Descriptive leaflets are available from the Director, SSRC Survey Archive, University of Essex, Wivenhoe Park, Colchester, Essex.

point of view of the sociologist, whose study centres on groups and relationships. This is one reason why official statistics are sometimes used more as background information in sociological studies rather than being central to the analysis. An exceptional field, though, is that of housing and household composition. For instance, the 1971 Census of Population gave detailed data on household and family types on a 10 per cent sample basis. Again, on the positive side, it must be strongly asserted that official sources provide data on large populations (of cities, counties and the nation as a whole) which would not be obtainable in any other way; and this enables certain useful kinds of analysis to be conducted, e.g. the comparative study of towns and cities or urban sub-areas (see Moser and Scott, 1961; Jones, 1960; Herbert and Johnston, 1978). Indeed, it can be confidently asserted that the sociographic research possibilities of sources such as the Census have only recently begun to be explored. There is little doubt that these and other areas of investigation would be more likely to be furthered if official sources were themselves better understood, and hence there is a need to provide elucidation here. The social scientist who is not a specialist in this field must know what material is available, its scope and limitations, and the ways in which it is properly to be interpreted.

In recent years it has become a good deal easier to find one's way around the main sources. Since its appearance in 1976, the basic reference has been the invaluable *Guide to Official Statistics* published by Her Majesty's Stationery Office (HMSO), a comprehensive volume covering all official and some non-official sources of statistics for the UK.* This publication superseded the earlier and much briefer *List of Principal Statistical Series and Publications* (HMSO, 1972, 1974). What the *Guide* does is to give the user 'a broad indication of whether the statistics he wants have been compiled and, if so, where they have been published' (1976, p. v). The intention is that the investigator will proceed to other publications which themselves contain appropriate statistical classifications. Updated versions of the *Guide* are expected to appear at approximately annual intervals, and the first of these became available in 1978. The *Guide* incorporates geographical, social and economic statistics, organised into a number of sections. The first section includes general sources of statistics and this is followed by area statistics, and population and vital statistics. The fourth section covers social statistics, and within this broad field one finds as well as general sources, detailed information on electoral statistics, civil and criminal proceedings, the police service, the National Health services, social security, superannuation, housing, education, entertainment, religion and miscellaneous community services and home affairs. In subsequent sections of the *Guide* information is provided for various economically significant categories: labour, agriculture, production industries, transport, distribution, public services, prices, national income and expenditure, public

*A brief pamphlet on official statistics is also available from the Central Statistical Office and the Central Office of Information, entitled *Government Statistics, a brief guide to sources.*

finance, financial and business institutions and overseas transactions. There is also a well-organised index and a comprehensive bibliography.

It is fair to say that official statistics are never static and an important quarterly publication to be considered in conjunction with the *Guide* is *Statistical News* (HMSO), which dates from 1968. This aims to provide a comprehensive account of current developments in UK official statistics and seeks to give general assistance to users. There are notes on the detailed changes and innovations which periodically take place, as well as articles on special topics. *Statistical News* is important to anyone who wishes to keep abreast of the very latest information.

Early in an inquiry the social scientist can also usefully refer to volumes of the series *Reviews of United Kingdom Statistical Sources*, edited by Professor W. F. Maunder (Pergamon Press). This series aims to cover all sources of economic and social statistics in the UK, and each review provides a guide to the location of both official and unofficial statistical information published over the last thirty years. Additional volumes continue to appear but among the existing ones dealing with the social field, one should note particularly *Volume II: Central Government Routine Health Statistics*, *Volume III: Housing in Great Britain and Northern Ireland*, *Volume IV: Leisure and Tourism* and *Volume V: General Sources of Statistics*.

After locating broadly relevant material, the social scientist will proceed to the publications which themselves contain data. These are of two main types. The first is produced by individual government departments which are the original source of the information. An example is the annual six-volume publication *Statistics of Education*, produced by the Department of Education and Science. The second type consists of summary publications (or 'digests') which bring together data in a particular field from a number of departmental sources. This latter type of publication has the advantage for the investigator that he is provided with an overview and is, therefore, unlikely to miss material because of departmental boundaries. The summary publication also lists the original sources, so that further inquiry is facilitated.

General Digests

A key annual publication presenting selected data in the social field is *Social Trends*. This has appeared since 1970 and aims to give a balanced picture of developments and patterns in society. Data are presented in such a way as to meet the needs of the policy-maker and administrator together with those of the general public in as straightforward and non-technical a style as possible. One is provided with important series of social and demographic statistics in colour charts and tables. In recent years the areas covered have included population, households and families, social groups, education, employment, income and wealth, resources and expenditure, health, housing, the environment, leisure, public safety and law enforcement. There are also articles on selected topics, usually including a commentary on some aspect of social change. (For instance, in *Social Trends*, no. 9, 1978, there appeared an article, 'Housing tenure in England and Wales: the present situation and recent trends', by A. E. Holmans of the Department of the Environment.)

A second basic publication which is broader in scope than *Social Trends*, since it incorporates many economic as well as social statistics, is the *Annual Abstract of Statistics*. This volume brings together information for the UK contributed by virtually every government department and also by other organisations. The figures are generally provided in such a way that comparison is possible over the previous ten years. Tables are grouped under various headings, and in 1977 the first five were: area and climate; population and vital statistics; social conditions; justice and crime; and education; there followed thirteen categories of predominantly economic significance. A publication with links to the *Annual Abstract* is the *Monthly Digest of Statistics*. The latter provides statistical series along somewhat similar lines to the former, but covering monthly or quarterly periods. The coverage of the *Monthly Digest* is, however, more limited than that of the *Annual Abstract* and it contains fewer tables. In the January issue of the *Monthly Digest* there appears an annual supplement of definitions and explanatory notes, applying not only to series in the periodical, but also to corresponding ones in the *Annual Abstract* and in *Regional Statistics* (see below).

Sometimes it is valuable to examine statistical information not simply for the UK (or Great Britain) as a whole, but for identifiable regions (or countries) within it. Some data of this kind appear in each of the general digests mentioned so far. For instance, *Social Trends* brings together various social statistics which are analysed by region and sometimes maps and charts are used for illustrative purposes. However, the most systematic coverage of this type is provided by *Regional Statistics*, which supplies various figures for the UK but also gives separate analyses for the eight regions of England, for Wales, for Scotland and for Northern Ireland. This annual publication includes sections on population, social services and education, and also gives data derived from income and household surveys (see below). Where it is desired to isolate data on Scotland, Wales or Northern Ireland and to scrutinise certain regional differences therein, the appropriate publication should be chosen from the (annual) *Scottish Abstract of Statistics*, the (annual) *Digest of Welsh Statistics*, or the (biannual) *Digest of Statistics, Northern Ireland*.

A general digest of a somewhat different kind, compiled by the Central Statistical Office and published by Penguin, is *Facts in Focus* (1972, further editions in 1974, 1975, 1978). This contains much useful data – in social and other areas – presented in such a way that the general reader, rather than the specialist alone, can make use of them. Its scope overlaps that of the *Annual Abstract* at certain points and it includes sections on population and vital statistics, health, social security, justice and law, housing, education and leisure. At the end of the book is an Index of Sources, which enables the reader to follow up topics of interest. *Facts in Focus* is a useful and easily obtainable starting-point for statistical investigation in the social field. A somewhat similar publication which aims to show social trends during the last twenty or thirty years is *Britain in Figures: A Handbook of Social Statistics* by Alan F. Sillitoe (Penguin, 1973). This book displays statistical series in graph and chart form, concerning such matters as population, social affairs, education, labour and the mass media, all of which are interpreted by an informative commentary.

If it is desired to compare UK statistics with those of other countries, a valuable annual publication is the *Statesman's Year-Book* (Macmillan). This contains a wealth of information on social and other topics collected on an international basis, while providing a useful introduction to other digests. However, United Nations publications are of fundamental importance in the field and one must consult particularly – in the sphere of population and vital statistics – the *Demographic Yearbook* (United Nations, New York), and also the *Statistical Yearbook* (UN, New York) which includes data on population, manpower, housing, health and education. There is also a monthly *Bulletin of Statistics*, but this is almost exclusively devoted to economic data. The specialised agencies of the United Nations also produce their own reports, particularly noteworthy being the *Statistical Yearbook* from UNESCO (Paris) covering education, culture and communication and the *Yearbook of Labour Statistics* from the International Labour Office (Geneva). In addition to the United Nations, the European Communities (EEC) produce several annual volumes of comparative statistics from their Statistical Office (in Luxembourg), including *Basic Statistics of the Community*, as well as *Social Statistics* which appears six times a year.

Key Primary Sources

Turning from digests to primary UK sources themselves, of particular importance are the censuses and surveys which cover a number of topics in the social field. Probably best known to members of the general public are the Censuses of Population, which have generally been conducted every ten years and separately for England and Wales, Scotland and Northern Ireland. They attempt to enumerate everyone in the country on a particular day (although there was in addition a 10 per cent 'sample Census' in 1966 in England and Wales, and also in Scotland).

The main aim of the Census is to facilitate the estimation and forecasting of population size and structure for the UK as a whole and also for local authority areas and counties. Official intercensal population estimates are derived from the Census results taken together with information from the registration of births and deaths and migration records. Clearly, accurate figures relating to the age and sex structure of the population are needed in connection with the provision of new schools, hospitals and a whole range of basic services.

Population censuses of Great Britain date back to 1801, which makes it possible to determine many extended statistical series and facilitates historical inquiry (see for instance, Anderson, 1971), but for analyses of present-day society the 1971 Census of Population is most relevant and it also happens to be the Census which has produced the most extensive tabulations. These latter can be broadly categorised into those which focus on persons and those which focus on households. In the first category, one finds persons classified by age, sex and marital status; country of birth; parents' countries of birth; address one and five years prior to the Census; educational qualifications; economic position; occupation at the time of the Census and one year previously; and (for women) age at marriage and number and spacing of children. In addition, in Wales and Scotland persons are classified according to their ability to speak Welsh and Gaelic, respectively. Households are analysed by size; tenure of dwelling and number of rooms occupied; availability of amenities (fixed bath or shower, hot-water supply and flush toilet); availability of car; family composition; number of earners and dependent children; and characteristics of chief economic supporter.

Publications from the Census mainly take the form of a series of *County Reports*, giving figures for each county, and a series of predominantly national reports, giving detailed information on the various topic areas. Some titles of national reports from the 1971 Census are *Age, Marital Condition and General Tables (Great Britain)*, *Country of Birth Tables (Great Britain)* and *Qualified Manpower Tables (Great Britain)*. In fact, the main publications only appeared over an extended period of time, and to speed up the flow of information provisional figures were given in *Preliminary Reports* and *Advance Analysis*. Some analyses and reports were also produced from a 10 per cent random sample of Census forms. Unpublished data from the Census can also be obtained subject to certain conditions being met. In this connection a particularly useful service is the provision of standard statistics for small sub-populations, e.g. enumeration districts, wards and civil parishes. The availability of these data both facilitates town and country planning, and enables academic researchers to analyse the internal structure of larger (e.g. urban) populations and areas.

One weakness in connection with the presentation of recent Census reports is worth noting. The last three decennial Censuses have witnessed a considerable increase in the production of published tables to a position where, in 1971, the number was vast (there were approximately 26,000 pages of tabulations). Many of the tables are inevitably complex since they involve cross classification of characteristics. The difficulty for the user is that virtually no commentary is provided. This is a pity, since interpretation by those possessing statistical skills but also familiar with the problems of designing and operating the Census would be invaluable. The 1951 and earlier Census reports tended to include informed interpretation of this kind and it is unfortunate that the practice has since changed. If large numbers of complicated tables are required to 'speak for themselves', there may be few 'listeners'.

Another important primary source of social statistics is the *General Household Survey*, which began operation in 1971. This is a continuing sample survey of private households which collects data simultaneously on behalf of several government departments. It is conducted by the Social Survey Division of the Office of Population Censuses and Surveys. About 15,000 households are contacted and all those persons aged 16 or over are interviewed. The main aim is to assess social conditions and assist social policy formulation and evaluation.

Certain topics are covered in the Survey on a continuous basis, while others are introduced from time to time. In the first category are basic items like housing, internal migration, employment, education, health and health services, and fertility, while in the second are to be found topics such as household theft (dealt with in 1972–3), leisure (1973) and hospital waiting-lists (1973). It is possible to relate variations on specially investigated items to certain key classificatory variables which have been covered continuously. These latter refer to household composition, socio-economic group (see below), industry classification, income and country of birth. Material is published annually in volumes such as the *General Household Survey 1973*, but unpublished material is sometimes made available with the consent of the relevant government department.*

A third source of statistics worthy of particular attention is the *Family Expenditure Survey*, which dates from 1957. This again is a continuing inquiry, having as its main purpose the gathering of information used in connection with the weighting pattern of the Index of Retail Prices, but it has other aims as well. For instance, it is used to investigate the extent to which members of various types of household contributed to the household income. Over 10,000 households are contacted and their members are interviewed about income and expenditure and also asked to record all payments during the fortnight following the interview.

The basic information collected is organised under three main headings: characteristics of households (including such items as age and sex composition, housing tenure and availability of telephone), household expenditure (including rent or mortgage, insurances, also detailed individual items) and household income and employment (including occupational earnings, investment income and social security benefits). Findings are made available annually in publications such as *Family Expenditure Survey 1974* but data from the survey appear elsewhere, e.g. in the *Annual Abstract of Statistics*. Unpublished material may also be obtained from the Department of Employment subject to certain undertakings.

Important Classifications

Official statistics may be used by the social scientist for comparative purposes and, in particular, in the analysis of trends. This is only possible

*More details are available from Social Survey Division, Office of Population Censuses and Surveys, St Catherine's House, 10 Kingsway, London, WC2 B6JP.

because the classifications used in the presentation of official data are precisely defined and standardised (often in conformity with internationally agreed criteria). This means, for instance, that an investigator who obtains a sample of (say) the employed population in a specified district (e.g. a local authority area) is in a good position to determine (e.g. from the Census of Population) whether or not his sample is broadly representative of the wider population in terms of its distribution by industry or occupation. In this connection, he will need to consult the basic classificatory systems which have been developed.

The *Standard Industrial Classification* is a system of classification of business units (generally termed 'establishments') according to industry. It is devised in such a way as to enable the industrial structure of the UK to be portrayed. The 'establishments' classified are not to be equated exactly with businesses themselves. An 'establishment' is basically the unit for which appropriate statistical records are provided; normally this is business premises at a particular address, or if not, it is a business embracing more than one address, or even two businesses or departments at the same address. The 1968 edition of the *Standard Industrial Classification* categorises industries into twenty-seven orders (including I, Agriculture, forestry and fishing; XIII, Textiles; XXVII, Public administration and defence), and these orders are further sub-classified into a total of 181 'Minimum List Headings', a finer classification reflecting an even more specific characterisation of the type of product or process. The *Standard Industrial Classification, Alphabetical List of Industries* is an index showing the Minimum List Heading for each specific industry.

Perhaps of even greater utility to the social scientist than the classification of persons by industry is their classification by type of employment. Of course, those in a given industry frequently perform quite different functions, and the same function (e.g. performance of clerical work) may be executed in very different industrial settings. For Census purposes, persons are classified in respect of their economically significant activities in four ways – according to their industry, occupation, employment status and economic position. The Standard Industrial Classification is used for the first of these, while details relating to the other three classifications appear in volumes such as the *Classification of Occupations 1970* (which generally appear immediately before a Census). The classification by employment status involves the major distinction between the 'employee' and the 'self-employed', together with various significant sub-divisions, e.g. in the former category, managers, foremen and supervisors, apprentices and family workers are distinguished. The classification by economic position divides the economically active (distinguishing those in employment or out of it on Census night) from the economically inactive (retired persons, students and others).

Of primary importance for social purposes is the classification of economically active persons by occupation. This is achieved in two ways: by 'social class' and by 'socio-economic group'. In the *Classification of*

Occupations, 1970, all occupations are classified into 27 orders and over 220 Unit Groups, which then build up into the two basic categorisations. There are five 'social classes': I, professional, etc. occupations; II, intermediate occupations; III, skilled occupations; IV, partly skilled occupations; V, unskilled occupations. We are told that the categories are 'homogeneous in relation to the basic criterion of the general standing within the community of the occupations concerned' (*Classification of Occupations, 1966*, p. x). Despite the longevity of the classification (it dates from 1911) and its widespread use (partly stemming from its simplicity), it raises a number of problems. One basic point is that there is insufficient empirical evidence to support the claim that the categories are either homogeneous in respect of the selected criterion, or systematically ranked with respect to each other. Hence it may not be justified to conduct statistical tests on samples so classified, as though ordinal scaling had been achieved. There is also the specific difficulty that social class III contains a half or more of the gainfully employed population, suggesting that for many purposes this rather artificial classification is a poor discriminator. Class III is also that category most widely felt to be inhomogeneous and it is common to distinguish within it between those in non-manual and those in manual employment. However, evidence is lacking regarding 'the standing within the community' of the resulting two sub-classes, so one continues to be in some doubt whether ordinal scaling has been achieved. It is perhaps better to treat the social class categorisation (and modifications of it) as a suggestive classification loosely related to considerations of status and prestige, but it cannot be viewed as an integral part of social class analysis as this is presented in the sociological literature.

The social scientist is on rather safer ground with the classification into seventeen socio-economic groups, for this does provide smaller and relatively homogeneous occupational groupings. In addition, there is no suggestion of prestige ranking. The number of groups may seem large and for research purposes unwieldly, but the social investigator can (and should) combine groups for particular types of analysis. It will also be found that in certain samples (e.g. those taken in specific localities) particular groups will be unrepresented. The socio-economic groups are: 1, employers and managers – large establishments; 2, employers and managers – small establishments; 3, professional workers – self-employed; 4, professional workers – employees; 5, intermediate non-manual workers; 6, junior non-manual employees; 7, personal service workers; 8, foremen and supervisors – manual; 9, skilled manual workers; 10, semi-skilled manual workers; 11, unskilled manual workers; 12, own-account workers (other than professional); 13, farmers – employers and managers; 14, farmers – own account; 15, agricultural workers; 16, members of armed forces; 17, occupations inadequately described.

A further occupational classification used by the Department of Employment was introduced in 1972. This appears as the *Classification of Occupations and Directory of Occupational Titles* and is published in three volumes. In this newer classification every attempt is made to group together occupations involving similar work tasks, so as to facilitate the production of more meaningful statistics relating to the mobility of workers between industries and occupations. Deriving from this classification is a *List of Key Occupations for Statistical Purposes*, which incorporates about 400 separate occupations under 18 major groups and is used as the basis for collecting and analysing statistical information obtained from firms.

Statistical Sources in Specialised Areas

The publications so far considered enable the social scientist to gain an accurate idea of the material available in a selected field and also to gain direct access to relevant data in digests and some key primary sources. To pursue a topic further will require scrutiny of more specialised sources generally produced by individual government departments. It is worthwhile indicating here some particularly important publications in the four main areas of social statistics: crime, health and welfare, housing and education.

Basic in the first field are the *Criminal Statistics, England and Wales*, *Criminal Statistics, Scotland* and the *Ulster Year Book*. In respect of criminal proceedings the data in the first two publications are comparable, if due allowance is made for the different judicial procedures applying. Both provide figures on persons found guilty in court proceedings shown by offence, sex, age group and sentence. The England and Wales publication also gives detailed data on offences recorded as known to the police and indicates the proportion cleared up. In all three publications summary figures are provided, enabling comparison to be made with previous years.

In the health field of particular importance are the *Health and Personal Social Service Statistics for England*, the *Health and Personal Social Service Statistics for Wales* and the *Scottish Health Statistics*. All three publications include data, generally relating to the latest five years, on notifications of infectious diseases, sexually transmitted diseases, cancer and abortions. Figures are also provided concerning National Health Service finance, the manpower situation and aspects of the dental services. As their names suggest, the first two publications also take in the personal social service field, including information on manpower, finance and the extent of the services provided, while broadly comparable data appear in *Scottish Social Work Statistics*. Basic in its field too is *Social Security Statistics*, which brings together for Great Britain (and to some extent for Northern Ireland) information on such matters as national insurance benefits and contributions, family allowances and supplementary benefits.

The single quarterly publication which provides the main series of data on housing in Great Britain (and a limited amount of information on Northern Ireland) is *Housing and Construction Statistics*. This includes coverage of house-building performance and gives details relating to local authority housing, house renovation grants, slum clearance and housing costs and prices. In addition, it is worth stressing that Census data are particularly extensive in this sphere and of particular relevance are *Census 1971: England and Wales: Housing Tables* (and corresponding publications for Scotland and Northern Ireland), *Census 1971: Great Britain: Housing Summary Tables*, as well as the various 1971 Census *County Reports*. These volumes contain considerable detail on such matters as type of household, rooms occupied, tenure and possession of amenities (hot water, fixed bath or shower, inside or outside flush toilet). Housing is also a topic covered regularly in the *General Household Survey*.

In education the main sources are *Statistics of Education*, which covers England and Wales and appears annually in six volumes (successively dealing with schools; school-leavers' CSE and GCE qualifications; further education; teachers; finance and awards; universities), taken together with *Scottish Educational Statistics* and *Northern Ireland Education Statistics*. In addition, *Education Statistics for the United Kingdom* brings together much information for the whole nation. These volumes cover such topics as public expenditure on education, numbers of pupils and schools, and teacher training. Education is also a topic which figures prominently in the Census (see, for instance, *Census 1971: Great Britain: Classified Manpower Tables*) and is dealt with on a regular basis in the *General Household Survey*.

Examples

Having sketched the general scope of official statistics, it may be worthwhile to exemplify their usefulness by indicating how the social investigator might proceed to assemble data on selected topics. This can serve to exhibit the contribution made by general digests as well as more specialised statistical volumes. The two chosen topics – one broad and one narrow – fall within the educational field. The first concerns sex differences in educational careers and qualifications. Let us suppose that the social investigator has the objective of assembling data which bear on the question of whether young people of each sex differ in their educational experiences as they complete their formal education and, if they do, whether the differences have become greater or smaller in the recent past. Since this inquiry is relevant among other things to the issue of whether males and females are equally equipped to play the same part in the sphere of paid work, it is germane to the general theme of the attainment of sexual equality.

Perusal of the general digests reveals that considerable data are available on this topic. *Social Trends*, the *Annual Abstract of Statistics* and *Facts in Focus*, each contain an extensive section on education, including many tables which distinguish men and women and also exhibit change over a period of time. Also *Social Trends* regularly contains commentaries on aspects of social change, and that in the 1977 volume is directly pertinent since it is entitled 'Fifteen to twenty-five: a decade of transition', and is produced by staff of the Central Statistical Office (*Social Trends*, 1977, pp. 10–25). The article is concerned with the relevant age group, but it is necessary for the investigator to be selective since references are made not simply to education, but also to such matters as family building and the place of young people in the labour force. In *Social Trends*, as in the other volumes, the reader is shown the source of the data, e.g. *Statistics of Education*, and he will proceed there for more detail.

In assembling tables for further analysis the investigator needs to keep in mind that there are three areas to consider – schools, further education and higher education. In examining sex differences in respect of schooling, a key issue concerns leaving qualifications. Are these similarly distributed for boys and girls? The relevant data for England and Wales are reproduced here as Table 11.1.

One observation which can be made is that for each of the years shown a greater percentage of boys than girls achieved the highest qualification – three or more A level passes – but for the later years the 'gap' was narrower. In fact, the ratio of the percentage of boys with this qualification to that for girls declined from 1.9 in 1965–6 to 1.4 in 1975–6. Indeed, over the ten-year period the rise in the proportion of girls achieving A level standard as a whole was quite substantial. At the other extreme, one can note the decline in those leaving school without any qualification. In this respect, there is a problem of interpretation since, for the earliest year, two categories are shown as amalgamated, but it is clear that the percentage decline for girls in those two categories over ten years has been greater than that experienced by boys. Summarising, one can say that over the period there was a discernible movement towards greater overall equality in the distribution of educational qualifications for boys and girls, but the position was complicated by the slight tendency, at any rate in more recent years, for male performance to be more variable. (For details of changes in the distribution by sex of subjects studied for these examinations, see, for instance, *Statistics of Education, 1976: Vol. 2, School Leavers C.S.E. and G.C.E.*, tables 28–30, pp. 48–50. There it is shown, for instance, that males continue to predominate among those who pass mathematics and physical science subjects, but over the period 1966–76 the proportion of girls among the successful candidates in these disciplines steadily increased.)

Social Trends also provides key tabulations for further and higher education (see Tables 11.2 and 11.3). Table 11.2 exhibits certain trends in further education, and one of the more striking is that the percentage of women among students in public-sector and assisted establishments in the UK rose from 36.9 per cent in 1966 to 47.5 per cent in 1975. (Unfortunately the figures for those in evening institutes are not distinguished by sex.) On the other hand, Table 11.3 shows that in the higher education sector as a whole in a recent ten-year period the number of women increased by 79.9 per cent, while the corresponding increase for men was 57.4 per cent. Hence there was a substantial sex difference in the rate of

Table 11.1 *Academic Attainment of School-Leavers*
England and Wales

Percentages

	1965–6		*1973–4*		*1974–5*[3]		*1975–6*[4]	
	Boys	*Girls*	*Boys*	*Girls*	*Boys*	*Girls*	*Boys*	*Girls*
3 or more A level passes	8.9	4.8	9.2	6.6	9.3	6.8	9.9	7.0
2 A level passes	4.0	3.8	4.1	4.4	4.0	4.2	4.1	4.5
1 A level pass	2.8	2.8	3.1	3.5	2.8	3.6	2.9	3.3
5 or more higher-grade O level GCE or CSE[1]	7.1	9.6	7.7	10.0	7.4	9.7	5.4	7.4
1–4 higher grades, GCE or CSE[1]	15.2	16.4	23.1	25.8	24.0	26.9	24.6	28.3
1 or more other grades, GCE or CSE[2]	62.0 (both rows)	62.5 (both rows)	30.7	29.7	31.9	30.4	34.2	32.5
No GCE or CSE qualifications			22.1	20.0	20.6	18.4	18.9	17.0
All leavers (thousands)	320.81	302.39	349.64	331.81	353.68	338.11	363.88	343.56

Notes:

[1] GCE O level grades, A–C (including O level passes on A level papers) and CSE grade 1.

[2] From 1974–5, GCE grades D and E are included.

[3] From 1974–5, GCE O level attainments were graded A–E and candidates with grades A–C are equated to the standard for the former pass.

[4] Provisional figures.

(*Source: Social Trends*, no. 8, 1977, table 4.7 p. 74.)

increase (tending to reduce overall sexual inequality), but as the commentary points out (*Social Trends*, no. 8, 1977, p. 14): 'Between 1965 and 1974 the increase in total numbers in the sector where females are predominant – colleges of education – was only 38 per cent while the increase in the sector where females are least well represented – further education advanced courses – was 139 per cent.' So the picture is in some respects complicated, but nevertheless the generalisation seems justified that there has been a steady movement towards greater overall equality in the distributions of educational qualifications among young males and females.

The second, altogether narrower, educational topic which can serve an illustrative purpose concerns the quality of the first degrees obtained in various academic subjects. It is a familiar fact that the first degrees which students obtain in universities are sometimes gained with, and sometimes without, honours and that honours degrees are classified into various classes, namely, first class, second class (frequently divided into an upper and lower division) and third class. The best performance is that indicated by the award of first-class honours. A question which may reasonably be asked is this: are students pursuing courses in different academic disciplines equally likely to achieve the highest standard?

Being narrow, this question is not to be answered by reference to general statistical digests, so the source to peruse is the *Statistics of Education, Vol. 6: Universities*. In the volume referring to 1975, information on the degrees and diplomas awarded appears in tables 18–24 (pp. 38–52); table 22 (pp. 44–7) classifies degrees according to their subject and quality. There it is shown that in total there were 54,114 initial degrees and 3,611 (or 6.7 per cent) were obtained with first-class honours. However, the proportion achieving this distinction varied considerably from one subject group and discipline to another. For instance, among the nine subject groups distinguished the proportion of first-class degrees was as high as 10.8 per cent in science but as low as 3.5 per cent in social, administrative and business studies, and 1.8 per cent in education (although in this latter field only a small proportion of students embarked on honours courses).

Some further simple calculations on the figures given enable the individual academic subjects themselves to be ranked according to the proportion of 'firsts'. At the top comes physiology and/or anatomy (with 16.7 per cent of 'firsts') and this is followed (among the major subjects) by physics (with 14.8 per cent); last (of the major subjects) is government and public administration (with 2.1 per cent). It can be shown that this particular ordering of subjects is by no means simply a temporary phenomenon of 1975, but exhibits a fair degree of stability during the decade of the 1970s. The conclusion which can be drawn is that the chances of students pursuing different courses obtaining a first-class degree are by no means equal. In fact, generally speaking a student is about three times as likely to achieve that standard in a pure science as compared with a social subject. Whether the explanation involves unequal entry requirements, characteristics of the

Table 11.2 *Further Education*

UK — Thousands

	UK 1966	UK 1971	UK 1974	UK 1975	England and Wales 1975	Scotland 1975	Northern Ireland 1975
Students in public-sector and assisted establishments[1]							
Major establishments:							
Full-time	207	282	328	481	440	29	12
Sandwich	23	43	48	52	47	4	1
Day release	693	661	625	616	541	61	14
Other part-time day	92	138	188	214	203	11	1
Evening only	854	792	828	847	802	31	14
Total	1,868	1,915	2,017	2,210	2,031	137	42
of which – men	1,178	1,144	1,104	1,161	1,049	89	23
– women	690	771	914	1,049	982	48	18
of which aged – 15–17[2]	546	498	467	460	400	44	16
18–20	515	476	473	557	499	47	11
21 and over	807	941	1,077	1,193	1,132	46	14
Evening institutes[3]							
Total	1,436	1,561	1,893	2,030	1,982	14	34
Total students in major establishments and evening institutes	3,304	3,476	3,910	4,239	4,013	151	75
of which: on courses leading to recognised qualifications:							
advanced[4]	177	222	249	369	338	26	5

non-advanced	1,064	1,078	1,048	1,105	963	110	32
on other courses	2,060	2,176	2,574	2,731	2,712	15	4
Students taking courses of adult education provided by responsible bodies[5,6]	224	255	269	275	275	–	–

Notes:

[1] Overseas students at UK establishments are included.
[2] In England and Wales, aged 16–17.
[3] Excluding students on non-vocational courses in Scotland.
[4] Courses leading to advanced-level qualifications, i.e. qualifications above the GCE A level, Scottish Certificate of Education 'H' grade, and Ordinary National Diploma or Ordinary National Certificate.
[5] Universities Workers' Educational Association and Welsh National Council of YMCA.
[6] England and Wales only.
(*Source: Social Trends*, no. 8, 1977, table 4.12, p. 77.)

Table 11.3 *Higher Education: Number and Age of Students*[1]

UK — Thousands

	Men					Women				
	1965–6	*1972–3*	*1973–4*	*1974–5*	*1975–6*	*1965–6*	*1972–3*	*1973–4*	*1974–5*	*1975–6*
Full-time students:										
Universities[1]	128.1	170.5	170.8	172.4	178.1	46.1	76.3	80.4	85.3	90.6
Colleges of Education[2,3,4]	24.2	37.5	38.8	33.1	4.0	61.0	89.7	91.5	85.3	10.3
Further-education advanced courses[3,4]	39.1	78.1	80.6	84.1	119.2	11.8	29.9	33.4	37.4	113.0
Total full-time students	191.4	286.1	290.2	289.6	301.3	118.9	195.9	205.3	208.0	213.9
of which:										
18 and under	–	30.2	29.9	29.9	31.7	–	32.5	32.1	30.0	30.4
19–20	–	101.8	102.7	101.8	104.3	–	87.8	89.4	90.3	89.9
21–24	–	105.0	106.7	105.1	108.6	–	48.8	54.8	56.3	59.8
25 and over	–	49.1	51.0	52.9	56.7	–	26.8	28.9	31.5	33.8
Part-time students:										
Universities	14.6	18.2	18.5	19.0	19.3	3.0	5.3	5.9	6.4	7.0
Further education: advanced courses										
Part-time day courses	56.9	68.5	69.7	76.5	80.4	3.1	8.7	10.6	12.9	15.5
Evening only courses	47.9	34.5	33.4	33.7	35.1	2.3	4.9	5.1	5.4	5.9
Total part-time students	119.4	121.2	121.6	129.2	134.8	8.4	18.9	21.6	24.7	28.4

Notes:

[1] Including overseas students.

[2] Students in university departments of education are included under universities.

[3] Students in art teacher training centres in further-education establishments and departments of education in polytechnics are included under further education.

[4] For 1975–6, Colleges of Education relate to Scotland and Northern Ireland only; figures for the former colleges of education for England and Wales are included in further education.

(*Sources: Social Trends*, no. 8, 1977, table 4.14, p. 78.)

disciplines themselves, or differing course structures, can only be revealed by further investigation.

Conclusion

A wealth of data is available in the form of official statistics. There are certain difficulties in using it of which two are particularly noteworthy. One problem (of a purely statistical kind!) is presented by the division of the nation into England and Wales, Scotland and Northern Ireland. Sometimes it is difficult to assemble statistics for the UK as a whole, since figures may be available in a certain form for England and Wales but those relating to Scotland and/or Northern Ireland are not strictly comparable. There may be good reasons for this (as when there are structural variations in the provision of services, for instance, those in the educational field), but occasionally it is simply an unfortunate administrative artefact. Certainly, the social investigator interested in the national picture must keep these divisions constantly in mind.

A further problem concerns comparative work on a defined population or area over extended periods of time. The difficulty here is that the categories used in tabulations, or their precise definitions (e.g. in the Censuses of Population), may change hence frustrating accurate comparison. In the production of official statistics over long periods there is inevitably tension between the need to keep abreast of social developments (and changing policy implications), and the desirability of standardising tabulations and categories in order to facilitate comparison. The only method of 'having it both ways' would be to increase the volume of data but this would be both expensive and eventually counterproductive, since users may be overwhelmed by sheer quantity. Given the compromises worked out in practice, the investigator must consult carefully both the numerous detailed footnotes which so often accompany tabulations (see Tables 11.1–11.3), and those publications concerned specifically with classification.

Despite these problems, the social scientist is generally in a strong position, since there has been a very real expansion in both the quantity and quality of information relating to social conditions in the UK in recent years. Official statistics are better organised and more accessible chiefly because of the co-ordinating function so effectively performed by the Central Statistical Office. Of course, the data which an investigator requires may not always be provided in the ideal form either because the information is insufficiently detailed, or the coverage of the data is too wide. In such cases it is necessary to make the best possible use of the available information. Sometimes it may be appropriate to approach a research topic indirectly with data gathered from various official and unofficial sources. Then the central problem can perhaps be tackled directly by means of interviewing or survey work. Without answering every problem, official statistics can at least provide a springboard for further inquiry.

References

Anderson, M. (1971), *Family Structure in Nineteenth Century Lancashire* (Cambridge: Cambridge University Press).
Benjamin, B. (1968), *Health and Vital Statistics* (London: Allen & Unwin).
Blalock, H. M. Jr (1972), *Social Statistics*, 2nd edn (London: McGraw-Hill).
Cochran, W. G. (1953), *Sampling Techniques* (London: Wiley).
Downey, K. J. (1975), *Elementary Social Statistics* (New York: Random House).
Edwards, B. (1974), *Sources of Social Statistics* (London: Heinemann).
Fraser, D. A. S. (1958), *Statistics: An Introduction* (New York: Wiley).
Guilford, J. P. (1956), *Fundamental Statistics in Psychology and Education*, 3rd edn (London: McGraw-Hill).
Haber, A. and Runyon, R. P. (1973), *General Statistics*, 2nd edn (London: Addison-Wesley).
Herbert, D. and Johnston, R. J. (eds) (1978), *Social Areas in Cities: Processes, Patterns and Problems* (Chichester: Wiley).
Jones E. (1960), *A Social Geography of Belfast* (London: Oxford University Press).
Maxwell, A. E. (1972), *Basic Statistics for Medical and Social Science Students* (London: Chapman & Hall).
Moser, C. A. and Kalton, G. (1971), *Survey Methods in Social Investigation*, 2nd edn (London: Heinemann).
Moser, C. A., and Scott, W. (1961), *British Towns: A Statistical Study of Their Social and Economic Differences* (Edinburgh: Oliver & Boyd).
Neave, H. R. (1978), *Statistics Tables for Mathematicians, Engineers, Economists and the Behavioural and Management Sciences* (London: Allen & Unwin).
Siegel, S. (1956), *Nonparametric Statistics for the Behavioral Sciences* (London: McGraw-Hill).
Yeomans, K. A. (1968), *Statistics for the Social Scientist: Vol. 1, Introductory Statistics; Vol. 2, Applied Statistics* (Harmondsworth: Penguin).

Appendix: Statistical Tables

Table A

Areas under the Standard Normal Curve

Shows area $\Phi(z)$ below a given z-value

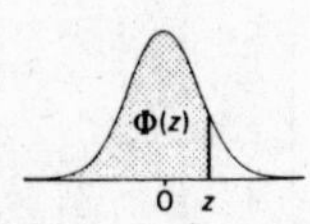

											SUBTRACT								
z	0	1	2	3	4	5	6	7	8	9	1	2	3	4	5	6	7	8	9
-3.9	0.0000	0000	0000	0000	0000	0000	0000	0000	0000	0000	0	0	0	0	0	0	0	0	0
-3.8	0.0001	0001	0001	0001	0001	0001	0001	0001	0001	0001	0	0	0	0	0	0	0	0	0
-3.7	0.0001	0001	0001	0001	0001	0001	0001	0001	0001	0001	0	0	0	0	0	0	0	0	0
-3.6	0.0002	0002	0001	0001	0001	0001	0001	0001	0001	0001	0	0	0	0	0	0	0	0	0
-3.5	0.0002	0002	0002	0002	0002	0002	0002	0002	0002	0002	0	0	0	0	0	0	0	0	0
-3.4	0.0003	0003	0003	0003	0003	0003	0003	0003	0003	0002	0	0	0	0	0	0	0	0	0
-3.3	0.0005	0005	0005	0004	0004	0004	0004	0004	0004	0003	0	0	0	0	0	0	0	0	0
-3.2	0.0007	0007	0006	0006	0006	0006	0006	0005	0005	0005	0	0	0	0	0	0	0	0	0
-3.1	0.0010	0009	0009	0009	0008	0008	0008	0008	0007	0007	0	0	0	0	0	0	0	0	0
-3.0	0.0013	0013	0013	0012	0012	0011	0011	0011	0010	0010	0	0	0	0	0	0	0	0	0
-2.9	0.0019	0018	0018	0017	0016	0016	0015	0015	0014	0014	0	0	0	0	0	0	0	0	0
-2.8	0.0026	0025	0024	0023	0023	0022	0021	0021	0020	0019	0	0	0	0	0	0	0	1	1
-2.7	0.0035	0034	0033	0032	0031	0030	0029	0028	0027	0026	0	0	0	0	0	1	1	1	1
-2.6	0.0047	0045	0044	0043	0041	0040	0039	0038	0037	0036	0	0	0	0	1	1	1	1	1
-2.5	0.0062	0060	0059	0057	0055	0054	0052	0051	0049	0048	0	0	0	1	1	1	1	1	1
-2.4	0.0082	0080	0078	0075	0073	0071	0069	0068	0066	0064	0	0	1	1	1	1	1	2	2
-2.3	0.0107	0104	0102	0099	0096	0094	0091	0089	0087	0084	0	1	1	1	1	2	2	2	2
-2.2	0.0139	0136	0132	0129	0125	0122	0119	0116	0113	0110	0	1	1	1	2	2	2	3	3
-2.1	0.0179	0174	0170	0166	0162	0158	0154	0150	0146	0143	0	1	1	2	2	2	3	3	4
-2.0	0.0228	0222	0217	0212	0207	0202	0197	0192	0188	0183	0	1	1	2	2	3	3	4	4
-1.9	0.0287	0281	0274	0268	0262	0256	0250	0244	0239	0233	1	1	2	2	3	4	4	5	5
-1.8	0.0359	0351	0344	0336	0329	0322	0314	0307	0301	0294	1	1	2	3	4	4	5	6	6
-1.7	0.0446	0436	0427	0418	0409	0401	0392	0384	0375	0367	1	2	3	3	4	5	6	7	8
-1.6	0.0548	0537	0526	0516	0505	0495	0485	0475	0465	0455	1	2	3	4	5	6	7	8	9
-1.5	0.0668	0655	0643	0630	0618	0606	0594	0582	0571	0559	1	2	4	5	6	7	8	10	11
-1.4	0.0808	0793	0778	0764	0749	0735	0721	0708	0694	0681	1	3	4	6	7	8	10	11	13
-1.3	0.0968	0951	0934	0918	0901	0885	0869	0853	0838	0823	2	3	5	6	8	10	11	13	14
-1.2	0.1151	1131	1112	1093	1075	1056	1038	1020	1003	0985	2	4	6	7	9	11	13	15	16
-1.1	0.1357	1335	1314	1292	1271	1251	1230	1210	1190	1170	2	4	6	8	10	12	14	16	19
-1.0	0.1587	1562	1539	1515	1492	1469	1446	1423	1401	1379	2	5	7	9	12	14	16	18	21
-0.9	0.1841	1814	1788	1762	1736	1711	1685	1660	1635	1611	3	5	8	10	13	15	18	20	23
-0.8	0.2119	2090	2061	2033	2005	1977	1949	1922	1894	1867	3	6	8	11	14	17	19	22	25
-0.7	0.2420	2389	2358	2327	2296	2266	2236	2206	2177	2148	3	6	9	12	15	18	21	24	27
-0.6	0.2743	2709	2676	2643	2611	2578	2546	2514	2483	2451	3	6	10	13	16	19	23	26	29
-0.5	0.3085	3050	3015	2981	2946	2912	2877	2843	2810	2776	3	7	10	14	17	21	24	27	31
-0.4	0.3446	3409	3372	3336	3300	3264	3228	3192	3156	3121	4	7	11	14	18	22	25	29	32
-0.3	0.3821	3783	3745	3707	3669	3632	3594	3557	3520	3483	4	8	11	15	19	23	26	30	34
-0.2	0.4207	4168	4129	4090	4052	4013	3974	3936	3897	3859	4	8	12	15	19	23	27	31	35
-0.1	0.4602	4562	4522	4483	4443	4404	4364	4325	4286	4247	4	8	12	16	20	24	28	32	35
-0.0	0.5000	4960	4920	4880	4840	4801	4761	4721	4681	4641	4	8	12	16	20	24	28	32	36
z	0	1	2	3	4	5	6	7	8	9	1	2	3	4	5	6	7	8	9

Table A-*continued*

z	0	1	2	3	4	5	6	7	8	9	ADD 1	2	3	4	5	6	7	8	9
0.0	0.5000	5040	5080	5120	5160	5199	5239	5279	5319	5359	4	8	12	16	20	24	28	32	36
0.1	0.5398	5438	5478	5517	5557	5596	5636	5675	5714	5753	4	8	12	16	20	24	28	32	35
0.2	0.5793	5832	5871	5910	5948	5987	6026	6064	6103	6141	4	8	12	15	19	23	27	31	35
0.3	0.6179	6217	6255	6293	6331	6368	6406	6443	6480	6517	4	8	11	15	19	23	26	30	34
0.4	0.6554	6591	6628	6664	6700	6736	6772	6808	6844	6879	4	7	11	14	18	22	25	29	32
0.5	0.6915	6950	6985	7019	7054	7088	7123	7157	7190	7224	3	7	10	14	17	21	24	27	31
0.6	0.7257	7291	7324	7357	7389	7422	7454	7486	7517	7549	3	6	10	13	16	19	23	26	29
0.7	0.7580	7611	7642	7673	7704	7734	7764	7794	7823	7852	3	6	9	12	15	18	21	24	27
0.8	0.7881	7910	7939	7967	7995	8023	8051	8078	8106	8133	3	6	8	11	14	17	19	22	25
0.9	0.8159	8186	8212	8238	8264	8289	8315	8340	8365	8389	3	5	8	10	13	15	18	20	23
1.0	0.8413	8438	8461	8485	8508	8531	8554	8577	8599	8621	2	5	7	9	12	14	16	18	21
1.1	0.8643	8665	8686	8708	8729	8749	8770	8790	8810	8830	2	4	6	8	10	12	14	16	19
1.2	0.8849	8869	8888	8907	8925	8944	8962	8980	8997	9015	2	4	6	7	9	11	13	15	16
1.3	0.9032	9049	9066	9082	9099	9115	9131	9147	9162	9177	2	3	5	6	8	10	11	13	14
1.4	0.9192	9207	9222	9236	9251	9265	9279	9292	9306	9319	1	3	4	6	7	8	10	11	13
1.5	0.9332	9345	9357	9370	9382	9394	9406	9418	9429	9441	1	2	4	5	6	7	8	10	11
1.6	0.9452	9463	9474	9484	9495	9505	9515	9525	9535	9545	1	2	3	4	5	6	7	8	9
1.7	0.9554	9564	9573	9582	9591	9599	9608	9616	9625	9633	1	2	3	3	4	5	6	7	8
1.8	0.9641	9649	9656	9664	9671	9678	9686	9693	9699	9706	1	1	2	3	4	4	5	6	6
1.9	0.9713	9719	9726	9732	9738	9744	9750	9756	9761	9767	1	1	2	2	3	4	4	5	5
2.0	0.9772	9778	9783	9788	9793	9798	9803	9808	9812	9817	0	1	1	2	2	3	3	4	4
2.1	0.9821	9826	9830	9834	9838	9842	9846	9850	9854	9857	0	1	1	2	2	2	3	3	4
2.2	0.9861	9864	9868	9871	9875	9878	9881	9884	9887	9890	0	1	1	1	2	2	2	3	3
2.3	0.9893	9896	9898	9901	9904	9906	9909	9911	9913	9916	0	1	1	1	1	2	2	2	2
2.4	0.9918	9920	9922	9925	9927	9929	9931	9932	9934	9936	0	0	1	1	1	1	1	2	2
2.5	0.9938	9940	9941	9943	9945	9946	9948	9949	9951	9952	0	0	0	1	1	1	1	1	1
2.6	0.9953	9955	9956	9957	9959	9960	9961	9962	9963	9964	0	0	0	0	1	1	1	1	1
2.7	0.9965	9966	9967	9968	9969	9970	9971	9972	9973	9974	0	0	0	0	0	1	1	1	1
2.8	0.9974	9975	9976	9977	9977	9978	9979	9979	9980	9981	0	0	0	0	0	0	0	1	1
2.9	0.9981	9982	9982	9983	9984	9984	9985	9985	9986	9986	0	0	0	0	0	0	0	0	0
z	0	1	2	3	4	5	6	7	8	9	1	2	3	4	5	6	7	8	9

z	0	1	2	3	4	5	6	7	8	9
3.0	0.998650	.998694	.998736	.998777	.998817	.998856	.998893	.998930	.998965	.998999
3.1	0.999032	.999065	.999096	.999126	.999155	.999184	.999211	.999238	.999264	.999289
3.2	0.999313	.999336	.999359	.999381	.999402	.999423	.999443	.999462	.999481	.999499
3.3	0.999517	.999534	.999550	.999566	.999581	.999596	.999610	.999624	.999638	.999651
3.4	0.999663	.999675	.999687	.999698	.999709	.999720	.999730	.999740	.999749	.999758
3.5	0.999767	.999776	.999784	.999792	.999800	.999807	.999815	.999822	.999828	.999835
3.6	0.999841	.999847	.999853	.999858	.999864	.999869	.999874	.999879	.999883	.999888
3.7	0.999892	.999896	.999900	.999904	.999908	.999912	.999915	.999918	.999922	.999925
3.8	0.999928	.999931	.999933	.999936	.999938	.999941	.999943	.999946	.999948	.999950
3.9	0.999952	.999954	.999956	.999958	.999959	.999961	.999963	.999964	.999966	.999967
4.0	0.999968	.999970	.999971	.999972	.999973	.999974	.999975	.999976	.999977	.999978
4.1	0.999979	.999980	.999981	.999982	.999983	.999983	.999984	.999985	.999985	.999986
4.2	0.999987	.999987	.999988	.999988	.999989	.999989	.999990	.999990	.999991	.999991
4.3	0.999991	.999992	.999992	.999993	.999993	.999993	.999993	.999994	.999994	.999994
4.4	0.999995	.999995	.999995	.999995	.999996	.999996	.999996	.999996	.999996	.999996
4.5	0.999997	.999997	.999997	.999997	.999997	.999997	.999997	.999998	.999998	.999998
4.6	0.999998	.999998	.999998	.999998	.999998	.999998	.999998	.999998	.999999	.999999
4.7	0.999999	.999999	.999999	.999999	.999999	.999999	.999999	.999999	.999999	.999999
4.8	0.999999	.999999	.999999	.999999	.999999	.999999	.999999	.999999	.999999	.999999
4.9	1.000000	1.00000	1.00000	1.00000	1.00000	1.00000	1.00000	1.00000	1.00000	1.00000

Table B
The Student t-distribution

Gives values of t which are exceeded with specified probability α when the number of degrees of freedom is ν

ν \ α	0.050	0.025	0.010	0.005	0.001	0.0005
1	6.314	12.71	31.82	63.66	318.3	636.6
2	2.920	4.303	6.965	9.925	22.33	31.60
3	2.353	3.182	4.541	5.841	10.21	12.92
4	2.132	2.776	3.747	4.604	7.173	8.610
5	2.015	2.571	3.365	4.032	5.893	6.869
6	1.943	2.447	3.143	3.707	5.208	5.959
7	1.895	2.365	2.998	3.499	4.785	5.408
8	1.860	2.306	2.896	3.355	4.501	5.041
9	1.833	2.262	2.821	3.250	4.297	4.781
10	1.812	2.228	2.764	3.169	4.144	4.587
11	1.796	2.201	2.718	3.106	4.025	4.437
12	1.782	2.179	2.681	3.055	3.930	4.318
13	1.771	2.160	2.650	3.012	3.852	4.221
14	1.761	2.145	2.624	2.977	3.787	4.140
15	1.753	2.131	2.602	2.947	3.733	4.073
16	1.746	2.120	2.583	2.921	3.686	4.015
17	1.740	2.110	2.567	2.898	3.646	3.965
18	1.734	2.101	2.552	2.878	3.610	3.922
19	1.729	2.093	2.539	2.861	3.579	3.883
20	1.725	2.086	2.528	2.845	3.552	3.850
21	1.721	2.080	2.518	2.831	3.527	3.819
22	1.717	2.074	2.508	2.819	3.505	3.792
23	1.714	2.069	2.500	2.807	3.485	3.768
24	1.711	2.064	2.492	2.797	3.467	3.745
25	1.708	2.060	2.485	2.787	3.450	3.725
26	1.706	2.056	2.479	2.779	3.435	3.707
27	1.703	2.052	2.473	2.771	3.421	3.690
28	1.701	2.048	2.467	2.763	3.408	3.674
29	1.699	2.045	2.462	2.756	3.396	3.659
30	1.697	2.042	2.457	2.750	3.385	3.646
31	1.696	2.040	2.453	2.744	3.375	3.633
32	1.694	2.037	2.449	2.738	3.365	3.622
33	1.692	2.035	2.445	2.733	3.356	3.611
34	1.691	2.032	2.441	2.728	3.348	3.601
35	1.690	2.030	2.438	2.724	3.340	3.591
36	1.688	2.028	2.434	2.719	3.333	3.582
37	1.687	2.026	2.431	2.715	3.326	3.574
38	1.686	2.024	2.429	2.712	3.319	3.566
39	1.685	2.023	2.426	2.708	3.313	3.558
40	1.684	2.021	2.423	2.704	3.307	3.551
45	1.679	2.014	2.412	2.690	3.281	3.520
50	1.676	2.009	2.403	2.678	3.261	3.496
60	1.671	2.000	2.390	2.660	3.232	3.460
70	1.667	1.994	2.381	2.648	3.211	3.435
80	1.664	1.990	2.374	2.639	3.195	3.416
90	1.662	1.987	2.368	2.632	3.183	3.402
100	1.660	1.984	2.364	2.626	3.174	3.390
120	1.658	1.980	2.358	2.617	3.160	3.373
150	1.655	1.976	2.351	2.609	3.145	3.357
∞	1.645	1.960	2.326	2.576	3.090	3.291

Table C
The χ^2 (Chi-square) Distribution

Gives values of χ^2 which are exceeded with specified probability α when the number of degrees of freedom is ν

ν \ α	0.050	0.025	0.010	0.005	0.001
1	3.841	5.024	6.635	7.879	10.827
2	5.991	7.378	9.210	10.597	13.815
3	7.815	9.348	11.345	12.838	16.268
4	9.488	11.143	13.277	14.860	18.465
5	11.070	12.832	15.086	16.750	20.517
6	12.592	14.449	16.812	18.548	22.457
7	14.067	16.013	18.475	20.278	24.322
8	15.507	17.535	20.090	21.955	26.125
9	16.919	19.023	21.666	23.589	27.877
10	18.307	20.483	23.209	25.188	29.588
11	19.675	21.920	24.725	26.757	31.264
12	21.026	23.337	26.217	28.300	32.909
13	22.362	24.736	27.688	29.819	34.528
14	23.685	26.119	29.141	31.319	36.123
15	24.996	27.488	30.578	32.801	37.697
16	26.296	28.845	32.000	34.267	39.252
17	27.587	30.191	33.409	35.718	40.790
18	28.869	31.526	34.805	37.156	42.312
19	30.144	32.852	36.191	38.582	43.820
20	31.410	34.170	37.566	39.997	45.315
21	32.671	35.479	38.932	41.401	46.797
22	33.924	36.781	40.289	42.796	48.268
23	35.172	38.076	41.638	44.181	49.728
24	36.415	39.364	42.980	45.558	51.179
25	37.652	40.646	44.314	46.928	52.620
26	38.885	41.923	45.642	48.290	54.052
27	40.113	43.194	46.963	49.645	55.476
28	41.337	44.461	48.278	50.993	56.893
29	42.557	45.722	49.588	52.336	58.302
30	43.773	46.979	50.892	53.672	59.703
35	49.802	53.204	57.342	60.274	66.619
40	55.758	59.342	63.691	66.766	73.402
45	61.657	65.410	69.957	73.166	80.077
50	67.505	71.420	76.154	79.490	86.661
55	73.311	77.380	82.292	85.749	93.168
60	79.082	83.298	88.379	91.952	99.607
65	84.821	89.177	94.422	98.105	105.998
70	90.531	95.023	100.425	104.215	112.317
75	96.217	100.839	106.393	110.286	118.599
80	101.879	106.629	112.329	116.321	124.839
85	107.522	112.393	118.236	122.325	131.041
90	113.145	118.136	124.116	128.299	137.208
95	118.752	123.858	129.973	134.247	143.344
100	124.342	129.561	135.807	140.169	149.449
120	146.567	152.212	158.951	163.649	173.618
150	179.580	185.801	193.207	198.360	209.265
200	233.996	241.058	249.446	255.264	267.541

5 Per Cent Points of the F-Distribution

Gives values of F which are exceeded with probability 0.05 in a one-tailed test when ν_1 and ν_2 represent the degrees of freedom associated with the larger and smaller variance estimates respectively

ν_2 \ ν_1 (α)	1	2	3	4	5	6	7	8	9	10	12	15	20	30	50	100	∞
1	161.4	199.5	215.7	224.6	230.2	234.0	236.8	238.9	240.5	241.9	243.9	245.9	248.0	250.1	251.8	253.0	254.3
2	18.51	19.00	19.16	19.25	19.30	19.33	19.35	19.37	19.38	19.40	19.41	19.43	19.45	19.46	19.48	19.49	19.50
3	10.13	9.552	9.277	9.117	9.013	8.941	8.887	8.845	8.812	8.786	8.745	8.703	8.660	8.617	8.581	8.554	8.526
4	7.709	6.944	6.591	6.388	6.256	6.163	6.094	6.041	5.999	5.964	5.912	5.858	5.803	5.746	5.699	5.664	5.628
5	6.608	5.786	5.409	5.192	5.050	4.950	4.876	4.818	4.772	4.735	4.678	4.619	4.558	4.496	4.444	4.405	4.365
6	5.987	5.143	4.757	4.534	4.387	4.284	4.207	4.147	4.099	4.060	4.000	3.938	3.874	3.808	3.754	3.712	3.669
7	5.591	4.737	4.347	3.120	3.972	3.866	3.787	3.726	3.677	3.637	3.575	3.511	3.445	3.376	3.319	3.275	3.230
8	5.318	4.459	4.066	3.838	3.687	3.581	3.500	3.438	3.388	3.347	3.284	3.218	3.150	3.079	3.020	2.975	2.928
9	5.117	4.256	3.863	3.633	3.482	3.374	3.293	3.230	3.179	3.137	3.073	3.006	2.936	2.864	2.803	2.756	2.707
10	4.965	4.103	3.708	3.478	3.326	3.217	3.135	3.072	3.020	2.978	2.913	2.845	2.774	2.700	2.637	2.588	2.538
11	4.844	3.982	3.587	3.357	3.204	3.095	3.012	2.948	2.896	2.854	2.788	2.719	2.646	2.570	2.507	2.457	2.404
12	4.747	3.885	3.490	3.259	3.106	2.996	2.913	2.849	2.796	2.753	2.687	2.617	2.544	2.466	2.401	2.350	2.296
13	4.667	3.806	3.411	3.179	3.025	2.915	2.832	2.767	2.714	2.671	2.604	2.533	2.459	2.380	2.314	2.261	2.206
14	4.600	3.739	3.344	3.112	2.958	2.848	2.764	2.699	2.646	2.602	2.534	2.463	2.388	2.308	2.241	2.187	2.131
15	4.543	3.682	3.287	3.056	2.901	2.790	2.707	2.641	2.588	2.544	2.475	2.403	2.328	2.247	2.178	2.123	2.066
16	4.494	3.634	3.239	3.007	2.852	2.741	2.657	2.591	2.538	2.494	2.425	2.352	2.276	2.194	2.124	2.068	2.010
17	4.451	3.592	3.197	2.965	2.810	2.699	2.614	2.548	2.494	2.450	2.381	2.308	2.230	2.148	2.077	2.020	1.960
18	4.414	3.555	3.160	2.928	2.773	2.661	2.577	2.510	2.456	2.412	2.342	2.269	2.191	2.107	2.035	1.978	1.917
19	4.381	3.522	3.127	2.895	2.740	2.628	2.544	2.477	2.423	2.378	2.308	2.234	2.155	2.071	1.999	1.940	1.878
20	4.351	3.493	3.098	2.866	2.711	2.599	2.514	2.447	2.393	2.348	2.278	2.203	2.124	2.039	1.966	1.907	1.843
21	4.325	3.467	3.072	2.840	2.685	2.573	2.488	2.420	2.366	2.321	2.250	2.176	2.096	2.010	1.936	1.876	1.812
22	4.301	3.443	3.049	2.817	2.661	2.549	2.464	2.397	2.342	2.297	2.226	2.151	2.071	1.984	1.909	1.849	1.783
23	4.279	3.422	3.028	2.796	2.640	2.528	2.442	2.375	2.320	2.275	2.204	2.128	2.048	1.961	1.885	1.823	1.757
24	4.260	3.403	3.009	2.776	2.621	2.508	2.423	2.355	2.300	2.255	2.183	2.108	2.027	1.939	1.863	1.800	1.733
25	4.242	3.385	2.991	2.759	2.603	2.490	2.405	2.337	2.282	2.236	2.165	2.089	2.007	1.919	1.842	1.779	1.711
26	4.225	3.369	2.975	2.743	2.587	2.474	2.388	2.321	2.265	2.220	2.148	2.072	1.990	1.901	1.823	1.760	1.691
27	4.210	3.354	2.960	2.728	2.572	2.459	2.373	2.305	2.250	2.204	2.132	2.056	1.974	1.884	1.806	1.742	1.672
28	4.196	3.340	2.947	2.714	2.558	2.445	2.359	2.291	2.236	2.190	2.118	2.041	1.959	1.869	1.790	1.725	1.654
29	4.183	3.328	2.934	2.701	2.545	2.432	2.346	2.278	2.223	2.177	2.104	2.027	1.945	1.854	1.775	1.710	1.638
30	4.171	3.316	2.922	2.690	2.534	2.421	2.334	2.266	2.211	2.165	2.092	2.015	1.932	1.841	1.761	1.695	1.622
35	4.121	3.267	2.874	2.641	2.485	2.372	2.285	2.217	2.161	2.114	2.041	1.963	1.878	1.786	1.703	1.635	1.558
40	4.085	3.232	2.839	2.606	2.449	2.336	2.249	2.180	2.124	2.077	2.003	1.924	1.839	1.744	1.660	1.589	1.509
45	4.057	3.204	2.812	2.579	2.422	2.308	2.221	2.152	2.096	2.049	1.974	1.895	1.808	1.713	1.626	1.554	1.470
50	4.034	3.183	2.790	2.557	2.400	2.286	2.199	2.130	2.073	2.026	1.952	1.871	1.784	1.687	1.599	1.525	1.438
55	4.016	3.165	2.773	2.540	2.383	2.269	2.181	2.112	2.055	2.008	1.933	1.852	1.764	1.666	1.577	1.501	1.411
60	4.001	3.150	2.758	2.525	2.368	2.254	2.167	2.097	2.040	1.993	1.917	1.836	1.748	1.649	1.559	1.481	1.389
65	3.989	3.138	2.746	2.513	2.356	2.242	2.154	2.084	2.027	1.980	1.904	1.823	1.734	1.635	1.543	1.464	1.370
70	3.978	3.128	2.736	2.503	2.346	2.231	2.143	2.074	2.017	1.969	1.893	1.812	1.722	1.622	1.530	1.450	1.353
75	3.968	3.119	2.727	2.494	2.337	2.222	2.134	2.064	2.007	1.959	1.884	1.802	1.712	1.611	1.518	1.437	1.338
80	3.960	3.111	2.719	2.486	2.329	2.214	2.126	2.056	1.999	1.951	1.875	1.793	1.703	1.602	1.508	1.426	1.325
85	3.953	3.104	2.712	2.479	2.322	2.207	2.119	2.049	1.992	1.944	1.868	1.786	1.695	1.593	1.499	1.416	1.313
90	3.947	3.098	2.706	2.473	2.316	2.201	2.113	2.043	1.986	1.938	1.861	1.779	1.688	1.586	1.491	1.407	1.302
95	3.941	3.092	2.712	2.467	2.310	2.195	2.108	2.037	1.980	1.932	1.855	1.773	1.682	1.579	1.484	1.399	1.292
100	3.936	3.087	2.696	2.463	2.305	2.191	2.103	2.032	1.975	1.927	1.850	1.768	1.676	1.573	1.477	1.392	1.283
120	3.920	3.072	2.680	2.447	2.290	2.175	2.087	2.016	1.959	1.910	1.834	1.750	1.659	1.554	1.457	1.369	1.254
150	3.904	3.056	2.665	2.432	2.274	2.160	2.071	2.001	1.943	1.894	1.817	1.734	1.641	1.535	1.436	1.345	1.223
200	3.888	3.041	2.650	2.417	2.259	2.144	2.056	1.985	1.927	1.878	1.801	1.717	1.623	1.516	1.415	1.321	1.189
∞	3.841	2.996	2.605	2.372	2.214	2.099	2.010	1.938	1.880	1.831	1.752	1.666	1.571	1.459	1.350	1.243	1.000

Table D - *continued*

1 Per Cent Points of the F-Distribution

Gives values of F which are exceeded with probability 0.01 in a one-tailed test when ν_1 and ν_2 represent the degrees of freedom associated with the larger and smaller variance estimates respectively

ν_2 \ ν_1	1	2	3	4	5	6	7	8	9	10	12	15	20	30	50	100	∞
1	4052	4999	5403	5625	5764	5859	5928	5981	6022	6056	6106	6157	6209	6261	6303	6334	6366
2	98.50	99.00	99.17	99.25	99.30	99.33	99.36	99.37	99.39	99.40	99.42	99.43	99.45	99.47	99.48	99.49	99.50
3	34.12	30.82	29.46	28.71	28.24	27.91	27.67	27.49	27.35	27.23	27.05	26.87	26.69	26.50	26.35	26.24	26.13
4	21.20	18.00	16.69	15.98	15.52	15.21	14.98	14.80	14.66	14.55	14.37	14.20	14.02	13.84	13.69	13.58	13.46
5	16.26	13.27	12.06	11.39	10.97	10.67	10.46	10.29	10.16	10.05	9.888	9.722	9.553	9.379	9.238	9.130	9.020
6	13.75	10.92	9.780	9.148	8.746	8.466	8.260	8.102	7.976	7.874	7.718	7.559	7.396	7.229	7.091	6.987	6.880
7	12.25	9.547	8.451	7.847	7.460	7.191	6.993	6.840	6.719	6.620	6.469	6.314	6.155	5.992	5.858	5.755	5.650
8	11.26	8.649	7.591	7.006	6.632	6.371	6.178	6.029	5.911	5.814	5.667	5.515	5.359	5.198	5.065	4.963	4.859
9	10.56	8.022	6.992	6.422	6.057	5.802	5.613	5.467	5.351	5.257	5.111	4.962	4.808	4.649	4.517	4.415	4.311
10	10.04	7.559	6.552	5.994	5.636	5.386	5.200	5.057	4.942	4.849	4.706	4.558	4.405	4.247	4.115	4.014	3.909
11	9.646	7.206	6.217	5.668	5.316	5.069	4.886	4.744	4.632	4.539	4.397	4.251	4.099	3.941	3.810	3.708	3.602
12	9.330	6.927	5.953	5.412	5.064	4.821	4.640	4.499	4.388	4.296	4.155	4.010	3.858	3.701	3.569	3.467	3.361
13	9.074	6.701	5.739	5.205	4.862	4.620	4.441	4.302	4.191	4.100	3.960	3.815	3.665	3.507	3.375	3.272	3.165
14	8.862	6.515	5.564	5.035	4.695	4.456	4.278	4.140	4.030	3.939	3.800	3.656	3.505	3.348	3.215	3.112	3.004
15	8.683	6.359	5.417	4.893	4.556	4.318	4.142	4.004	3.895	3.805	3.666	3.522	3.372	3.214	3.081	2.977	2.868
16	8.531	6.226	5.292	4.773	4.437	4.202	4.026	3.890	3.780	3.691	3.553	3.409	3.259	3.101	2.967	2.863	2.753
17	8.400	6.112	5.185	4.669	4.336	4.102	3.927	3.791	3.682	3.593	3.455	3.312	3.162	3.003	2.869	2.764	2.653
18	8.285	6.013	5.092	4.579	4.248	4.015	3.841	3.705	3.597	3.508	3.371	3.227	3.077	2.919	2.784	2.678	2.566
19	8.185	5.926	5.010	4.500	4.171	3.939	3.765	3.631	3.523	3.434	3.297	3.153	3.003	2.844	2.709	2.602	2.489
20	8.096	5.849	4.938	4.431	4.103	3.871	3.699	3.564	3.457	3.368	3.231	3.088	2.938	2.778	2.643	2.535	2.421
21	8.017	5.780	4.874	4.369	4.042	3.812	3.640	3.506	3.398	3.310	3.173	3.030	2.880	2.720	2.584	2.475	2.360
22	7.945	5.719	4.817	4.313	3.988	3.758	3.587	3.453	3.346	3.258	3.121	2.978	2.827	2.667	2.531	2.422	2.305
23	7.881	5.664	4.765	4.264	3.939	3.710	3.539	3.406	3.299	3.211	3.074	2.931	2.781	2.620	2.483	2.373	2.256
24	7.823	5.614	4.718	4.218	3.895	3.667	3.496	3.363	3.256	3.168	3.032	2.889	2.738	2.577	2.440	2.329	2.211
25	7.770	5.568	4.675	4.177	3.855	3.627	3.457	3.324	3.217	3.129	2.993	2.850	2.699	2.538	2.400	2.289	2.169
26	7.721	5.526	4.637	4.140	3.818	3.591	3.421	3.288	3.182	3.094	2.958	2.815	2.664	2.503	2.364	2.252	2.131
27	7.677	5.488	4.601	4.106	3.785	3.558	3.388	3.256	3.149	3.062	2.926	2.783	2.632	2.470	2.330	2.218	2.097
28	7.636	5.453	4.568	4.074	3.754	3.528	3.358	3.226	3.120	3.032	2.896	2.753	2.602	2.440	2.300	2.187	2.064
29	7.598	5.420	4.538	4.045	3.725	3.499	3.330	3.198	3.092	3.005	2.868	2.726	2.574	2.412	2.271	2.158	2.034
30	7.562	5.390	4.510	4.018	3.699	3.473	3.304	3.173	3.067	2.979	2.843	2.700	2.549	2.386	2.245	2.131	2.006
35	7.419	5.268	4.396	3.908	3.592	3.368	3.200	3.069	2.963	2.876	2.740	2.597	2.445	2.281	2.137	2.020	1.891
40	7.314	5.170	4.313	3.828	3.514	3.291	3.124	2.993	2.888	2.801	2.665	2.522	2.369	2.203	2.058	1.938	1.805
45	7.234	5.110	4.249	3.767	3.454	3.232	3.066	2.935	2.830	2.743	2.608	2.464	2.311	2.144	1.997	1.875	1.737
50	7.171	5.057	4.199	3.720	3.408	3.186	3.020	2.890	2.785	2.698	2.562	2.419	2.265	2.098	1.949	1.825	1.683
55	7.119	5.013	4.159	3.681	3.370	3.149	2.983	2.853	2.748	2.662	2.526	2.382	2.228	2.060	1.910	1.784	1.638
60	7.077	4.977	4.126	3.649	3.339	3.119	2.953	2.823	2.718	2.632	2.496	2.352	2.198	2.028	1.877	1.749	1.601
65	7.042	4.947	4.098	3.622	3.313	3.093	2.928	2.798	2.693	2.607	2.471	2.327	2.172	2.002	1.850	1.720	1.568
70	7.011	4.922	4.074	3.600	3.291	3.071	2.906	2.777	2.672	2.585	2.450	2.306	2.150	1.980	1.826	1.695	1.540
75	6.985	4.900	4.054	3.580	3.272	3.052	2.887	2.758	2.653	2.567	2.431	2.287	2.132	1.960	1.806	1.674	1.516
80	6.963	4.881	4.036	3.563	3.255	3.036	2.871	2.742	2.637	2.551	2.415	2.271	2.115	1.944	1.788	1.655	1.494
85	6.942	4.864	4.020	3.548	3.240	3.021	2.857	2.728	2.623	2.537	2.401	2.257	2.101	1.929	1.773	1.638	1.475
90	6.925	4.849	4.007	3.535	3.228	3.009	2.845	2.715	2.611	2.524	2.389	2.244	2.088	1.916	1.759	1.623	1.457
95	6.909	4.836	3.995	3.523	3.216	2.997	2.833	2.704	2.600	2.513	2.378	2.233	2.077	1.904	1.746	1.610	1.442
100	6.895	4.824	3.984	3.513	3.206	2.988	2.823	2.694	2.590	2.503	2.368	2.223	2.067	1.893	1.735	1.598	1.427
120	6.851	4.787	3.949	3.480	3.174	2.956	2.792	2.663	2.559	2.472	2.336	2.192	2.035	1.860	1.700	1.559	1.381
150	6.807	4.749	3.915	3.447	3.142	2.924	2.761	2.632	2.528	2.441	2.305	2.160	2.003	1.827	1.665	1.520	1.331
200	6.763	4.713	3.881	3.414	3.110	2.893	2.730	2.601	2.497	2.411	2.275	2.129	1.971	1.794	1.629	1.481	1.279
[illegible]	[illegible]	[illegible]	[illegible]	[illegible]	[illegible]	[illegible]	[illegible]	[illegible]	[illegible]	[illegible]	[illegible]	[illegible]	[illegible]	[illegible]	[illegible]	[illegible]	[illegible]

Table E

Mann-Whitney Rank-Sum Two-sample Test

Critical region: $U \leqslant$ tabulated value

n_L (1%) \ n_L (5%)	2	3	4	5	6	7	8	9	10	11	12	13	14	15	16	17	18	19	20	21	22	23	24	25	n_S (5%)
2		-	-	-	-	-	0	0	0	0	1	1	1	1	1	2	2	2	2	3	3	3	3	3	2
3	-		-	0	1	1	2	2	3	3	4	4	5	5	6	6	7	7	8	8	9	9	10	10	3
4	-	-		1	2	3	4	4	5	6	7	8	9	10	11	11	12	13	14	15	16	17	17	18	4
5	-	-	-		3	5	6	7	8	9	11	12	13	14	15	17	18	19	20	22	23	24	25	27	5
6	-	-	0	1		6	8	10	11	13	14	16	17	19	21	22	24	25	27	29	30	32	33	35	6
7	-	-	0	1	3		10	12	14	16	18	20	22	24	26	28	30	32	34	36	38	40	42	44	7
8	-	-	1	2	4	6		15	17	19	22	24	26	29	31	34	36	38	41	43	45	48	50	53	8
9	-	0	1	3	5	7	9		20	23	26	28	31	34	37	39	42	45	48	50	53	56	59	62	9
10	-	0	2	4	6	9	11	13		26	29	33	36	39	42	45	48	52	55	58	61	64	67	71	10
11	-	0	2	5	7	10	13	16	18		33	37	40	44	47	51	55	58	62	65	69	73	76	80	11
12	-	1	3	6	9	12	15	18	21	24		41	45	49	53	57	61	65	69	73	77	81	85	89	12
13	-	1	3	7	10	13	17	20	24	27	31		50	54	59	63	67	72	76	80	85	89	94	98	13
14	-	1	4	7	11	15	18	22	26	30	34	38		59	64	69	74	78	83	88	93	98	102	107	14
15	-	2	5	8	12	16	20	24	29	33	37	42	46		70	75	80	85	90	96	101	106	111	117	15
16	-	2	5	9	13	18	22	27	31	36	41	45	50	55		81	86	92	98	103	109	115	120	126	16
17	-	2	6	10	15	19	24	29	34	39	44	49	54	60	65		93	99	105	111	117	123	129	135	17
18	-	2	6	11	16	21	26	31	37	42	47	53	58	64	70	75		106	112	119	125	132	138	145	18
19	0	3	7	12	17	22	28	33	39	45	51	57	63	69	74	81	87		119	126	133	140	147	154	19
20	0	3	8	13	18	24	30	36	42	48	54	60	67	73	79	86	92	99		134	141	149	156	163	20
21	0	3	8	14	19	25	32	38	44	51	58	64	71	78	84	91	98	105	112		150	157	165	173	21
22	0	4	9	14	21	27	34	40	47	54	61	68	75	82	89	96	104	111	118	125		166	174	182	22
23	0	4	9	15	22	29	35	43	50	57	64	72	79	87	94	102	109	117	125	132	140		183	192	23
24	0	4	10	16	23	30	37	45	52	60	68	75	83	91	99	107	115	123	131	139	147	155		201	24
25	0	5	10	17	24	32	39	47	55	63	71	79	87	96	104	112	121	129	138	146	155	163	172		25
n_S (1%)	2	3	4	5	6	7	8	9	10	11	12	13	14	15	16	17	18	19	20	21	22	23	24	25	

equal sample sizes

n	1	2	3	4	5	6	7	8	9	10	11	12	13	14	15	16	17	18	19	20	21	22	23	24	25
5%	-	-	-	0	2	5	8	13	17	23	30	37	45	55	64	75	87	99	113	127	142	158	175	192	211
1%	-	-	-	-	0	2	4	7	11	16	21	27	34	42	51	60	70	81	93	105	118	133	148	164	180

n	26	27	28	29	30	31	32	33	34	35	36	37	38	39	40	41	42	43	44	45	46	47	48	49	50
5%	230	250	272	294	317	341	365	391	418	445	473	503	533	564	596	628	662	697	732	769	806	845	884	924	965
1%	198	216	235	255	276	298	321	344	369	394	420	447	475	504	533	564	595	627	660	694	729	765	802	839	877

Table F

Kolmogorov-Smirnov Two-sample Test

Critical region: $D \geqslant$ tabulated value$/(n_S n_L)$

Upper right triangle: 5% (rows n_S, columns n_L); lower left triangle: 1% (rows n_L, columns n_S).

1%: n_L	2	3	4	5	6	7	8	9	10	11	12	13	14	15	16	17	18	19	20	21	22	23	24	25	5%: n_S
2		-	-	-	-	-	16	18	20	22	24	26	26	28	30	32	34	36	38	38	40	42	44	46	2
3	-		-	15	18	21	21	24	27	30	30	33	36	36	39	42	45	45	48	51	51	54	57	60	3
4	-	-		20	20	24	28	28	30	33	36	39	42	44	48	48	50	53	60	59	62	64	68	68	4
5	-	-	-		24	28	30	35	40	39	43	45	46	55	54	55	60	61	65	69	70	72	76	80	5
6	-	-	24	30		30	34	39	40	43	48	52	54	57	60	62	72	70	72	75	78	80	90	88	6
7	-	-	28	35	36		40	42	46	48	53	56	63	62	64	68	72	76	79	91	84	89	92	97	7
8	-	-	32	35	40	48		46	48	53	60	62	64	67	80	77	80	82	88	89	94	98	104	104	8
9	-	27	36	40	45	49	55		53	59	63	65	70	75	78	82	90	89	93	99	101	106	111	114	9
10	-	30	36	45	48	53	60	63		60	66	70	74	80	84	89	92	94	110	105	108	114	118	125	10
11	-	33	40	45	54	59	64	70	77		72	75	82	84	89	93	97	102	107	112	121	119	124	129	11
12	-	36	44	50	60	60	68	75	80	86		81	86	93	96	100	108	108	116	120	124	125	144	138	12
13	-	39	48	52	60	65	72	78	84	91	95		89	96	101	105	110	114	120	126	130	135	140	145	13
14	-	42	48	56	64	77	76	84	90	96	104	104		98	106	111	116	121	126	140	138	142	146	150	14
15	-	42	52	60	69	75	81	90	100	102	108	115	123		114	116	123	127	135	138	144	149	156	160	15
16	-	45	56	64	72	77	88	94	100	106	116	121	126	133		124	128	133	140	145	150	157	168	167	16
17	-	48	60	68	73	84	88	99	106	110	119	127	134	142	143		133	141	146	151	157	163	168	173	17
18	-	51	60	70	84	87	94	108	108	118	126	131	140	147	154	164		142	152	159	164	170	180	180	18
19	38	54	64	71	83	91	98	107	113	122	130	138	148	152	160	166	176		160	163	169	177	183	187	19
20	40	57	68	80	88	93	104	111	130	127	140	143	152	160	168	175	182	187		173	176	184	192	200	20
21	42	57	72	80	90	105	107	117	126	134	141	150	161	168	173	180	189	199	199		183	189	198	202	21
22	44	60	72	83	92	103	112	122	130	143	148	156	164	173	180	187	196	204	212	223		194	204	209	22
23	46	63	76	87	97	108	115	126	137	142	149	161	170	179	187	196	204	209	219	227	237		205	216	23
24	48	66	80	90	102	112	128	132	140	150	168	166	176	186	200	203	216	218	228	237	242	249		225	24
25	50	69	84	95	107	115	125	135	150	154	165	172	182	195	199	207	216	224	235	244	250	262	262		25
1%: n_L / n_S	2	3	4	5	6	7	8	9	10	11	12	13	14	15	16	17	18	19	20	21	22	23	24	25	

equal sample **sizes**, critical region: $D \geqslant$ tabulated **value**/n

n	3	4	5	6	7	8,9	10-12	13	14,15	16,17	18,19	20-22	23	24-27	28-32	33-37	38-39
5%	-	4	5	5	6	6	7	7	8	8	9	9	10	10	11	12	12
1%	-	-	5	6	6	7	8	9	9	10	10	11	11	12	13	14	15

n	40-42	43-45	46-48	49-53	54	55-61	62-68	69	70-75	76-78	79-83	84-87	88-91	92-97	98-100
5%	13	13	14	14	15	15	16	16	17	17	18	18	19	19	20
1%	15	16	16	17	17	18	19	20	20	21	21	22	22	23	23

Table G
The Kruskal-Wallis Test

Critical region: $H \geqslant$ tabulated value

$k = 3$ sample sizes	5%	1%
2 2 2	-	-
3 2 1	-	-
3 2 2	4.714	-
3 3 1	5.143	-
3 3 2	5.361	-
3 3 3	5.600	7.200
4 2 1	-	-
4 2 2	5.333	-
4 3 1	5.208	-
4 3 2	5.444	6.444
4 3 3	5.791	6.745
4 4 1	4.967	6.667
4 4 2	5.455	7.036
4 4 3	5.598	7.144
4 4 4	5.692	7.654
5 2 1	5.000	-
5 2 2	5.160	6.533
5 3 1	4.960	-
5 3 2	5.251	6.909
5 3 3	5.648	7.079
5 4 1	4.985	6.955
5 4 2	5.273	7.205
5 4 3	5.656	7.445
5 4 4	5.657	7.760
5 5 1	5.127	7.309
5 5 2	5.338	7.338

$k = 3$ sample sizes	5%	1%
5 5 3	5.705	7.578
5 5 4	5.666	7.823
5 5 5	5.780	8.000
6 1 1	-	-
6 2 1	4.822	-
6 2 2	5.345	6.655
6 3 1	4.855	6.873
6 3 2	5.348	6.970
6 3 3	5.615	7.410
6 4 1	4.947	7.106
6 4 2	5.340	7.340
6 4 3	5.610	7.500
6 4 4	5.681	7.795
6 5 1	4.990	7.182
6 5 2	5.338	7.376
6 5 3	5.602	7.590
6 5 4	5.661	7.936
6 5 5	5.729	8.028
6 6 1	4.945	7.121
6 6 2	5.410	7.467
6 6 3	5.625	7.725
6 6 4	5.724	8.000
6 6 5	5.765	8.124
6 6 6	5.801	8.222
7 7 7	5.819	8.378
8 8 8	5.805	8.465

$k = 4$ sample sizes	5%	1%
2 2 1 1	-	-
2 2 2 1	5.679	-
2 2 2 2	6.167	6.667
3 1 1 1	-	-
3 2 1 1	-	-
3 2 2 1	5.833	-
3 2 2 2	6.333	7.133
3 3 1 1	6.333	-
3 3 2 1	6.244	7.200
3 3 2 2	6.527	7.636
3 3 3 1	6.600	7.400
3 3 3 2	6.727	8.015
3 3 3 3	7.000	8.538
4 1 1 1	-	-
4 2 1 1	5.833	-
4 2 2 1	6.133	7.000
4 2 2 2	6.545	7.391
4 3 1 1	6.178	7.067
4 3 2 1	6.309	7.455
4 3 2 2	6.621	7.871
4 3 3 1	6.545	7.758
4 3 3 2	6.795	8.333
4 3 3 3	6.984	8.659
4 4 1 1	5.945	7.909
4 4 2 1	6.386	7.909
4 4 2 2	6.731	8.346
4 4 3 1	6.635	8.231
4 4 3 2	6.874	8.621
4 4 3 3	7.038	8.876
4 4 4 1	6.725	8.588
4 4 4 2	6.957	8.871
4 4 4 3	7.142	9.075
4 4 4 4	7.235	9.287

$k = 5$ sample sizes	5%	1%
2 2 1 1 1	-	-
2 2 2 1 1	6.750	-
2 2 2 2 1	7.133	7.533
2 2 2 2 2	7.418	8.291
3 1 1 1 1	-	-
3 2 1 1 1	6.583	-
3 2 2 1 1	6.800	7.600
3 2 2 2 1	7.309	8.127
3 2 2 2 2	7.682	8.682
3 3 1 1 1	7.111	-
3 3 2 1 1	7.200	8.073
3 3 2 2 1	7.591	8.576
3 3 2 2 2	7.910	9.115
3 3 3 1 1	7.576	8.424
3 3 3 2 1	7.769	9.051
3 3 3 2 2	8.044	9.505
3 3 3 3 1	8.000	9.451
3 3 3 3 2	8.200	9.876
3 3 3 3 3	8.333	10.20

Table H
Random Numbers

49407	19320	03550	95272	89725	04148	80130	27441	58580	48974
29454	93427	48261	33970	27071	41907	52706	26593	56177	70961
25241	66874	75246	69296	60242	10703	12969	90648	75123	47334
02468	90461	89688	91631	91835	20886	63450	37902	68128	71870
69435	10988	67277	96170	77724	78311	15838	53105	88912	78049
40342	86089	07976	72803	63370	98477	52924	77511	19038	40305
91255	30062	82209	78496	45723	26690	00197	01044	40031	23958
19640	97421	89294	24165	40112	73259	70546	58773	80740	15247
97094	37081	42008	34675	73882	74429	27116	18743	52110	06017
44613	49506	93155	55264	90651	38978	83045	41652	89599	77686
68305	35793	23344	28616	52713	07480	50787	27434	38221	09948
85415	43422	18769	22256	60683	56850	75625	53832	73379	25066
95693	88060	40967	49214	83601	10928	13995	01602	78549	65636
51346	35573	73540	44047	81894	97881	98808	07475	82682	39229
29876	91257	67916	55543	04910	54817	00064	11451	55778	15845
23020	91127	36492	49725	25101	38574	16161	70688	64955	97762
20387	42900	70816	09288	36859	80581	99495	20911	49061	59712
70890	24023	98511	78737	92681	18906	44174	60213	72380	67657
24719	89153	26018	82284	02873	50138	97175	79744	92793	67267
22921	57406	34030	97165	28956	65216	47434	83581	54885	34756
27738	07626	68035	49093	56227	34705	85405	33130	88365	66475
94921	41465	88054	26802	61458	43285	11689	14651	52525	22821
84949	58541	15932	64469	69237	59690	07031	15189	42685	12402
31190	64902	16912	61425	63940	69846	83678	88599	60735	16353
22938	38795	57675	92614	04405	18713	24096	73716	49312	44148
10177	81086	44915	73573	41165	11812	24337	40865	73062	17749
26053	79260	14325	65134	07943	57583	69827	85215	56432	02495
95082	72403	98532	83049	15836	83460	30189	97616	16193	60260
20885	56027	98093	43285	06455	48510	53487	85596	32093	31132
12006	89103	22857	76849	24727	75505	61276	04354	15836	55361
48322	76947	61614	20459	79430	86711	68039	86197	23534	25641
98013	82682	34326	44489	64957	59567	54503	66436	05021	36255
09334	51835	13640	76488	83456	66906	27952	86200	81349	56820
40622	74324	28342	84970	06806	17654	25393	45749	15569	49523
20913	83062	96705	40214	52056	95595	69641	80264	68331	26242
20419	73850	62057	21476	58874	45093	43937	85659	80965	86236
58718	89349	42409	23730	07915	50959	49786	28670	22608	93331
21945	16088	80630	28848	80925	95582	33304	07532	86539	39792
10648	73413	33367	26591	83909	19570	88242	23258	67152	38067
33987	44662	10755	35927	11816	30216	69955	36958	05230	61272
65286	98315	01411	99018	46284	70624	43041	35415	83567	64596
46947	51250	72969	37996	41991	73271	56178	92838	21596	72495
91449	12740	12011	98376	96742	44382	32703	18017	33062	40518
23027	66240	06150	97386	20910	36959	57027	78806	26894	15626
77190	53484	51914	23909	65376	06566	35465	67021	81630	66003
19643	41964	57865	01946	51418	43129	38421	67691	68517	38828
71783	39109	62806	97147	96314	43100	98926	43094	89934	72274
65238	41996	87598	63589	89270	30007	00299	27177	65256	08929
92021	22580	52235	93505	02123	48807	88498	95866	45224	72248
95688	07898	22442	79188	37275	18053	25742	40488	63587	65830

70179	37665	74824	95559	09029	79382	98786	81667	73451	94196
84276	45611	69280	58659	29163	42366	83434	16564	47721	00007
79299	48337	76265	18064	83976	38367	16444	32212	22485	40246
74183	53308	07723	62211	52945	23806	98677	35496	20190	48981
85381	69831	10219	78203	80374	04039	67458	19358	40377	00157
17117	44369	49162	59855	96873	39687	30813	62663	89131	83424
07684	16102	99805	55435	30849	53433	57801	74434	14033	16601
13181	61384	16226	20043	05053	17315	87150	56240	67144	54462
33376	59377	16899	37918	52607	08931	77959	26295	58069	95907
44472	06322	88370	61791	35162	66388	51438	71281	42660	27210
10452	14399	78486	49513	90280	99587	32234	19021	86748	78215
72222	03569	83056	17483	29650	78357	78404	20591	91717	94585
19663	03430	15737	11384	21171	71898	06365	03372	97719	00206
17633	72800	24507	87554	52741	06868	52735	29142	30889	28776
85415	43422	18769	22256	60683	56850	67992	06739	81626	59453
70688	16161	38574	91127	23020	64955	97762	23909	74196	75423
55079	61532	60766	89325	65393	12744	05731	37897	23658	70390
18667	71760	00914	89301	50628	67695	00249	59302	41336	29363
66871	51598	03084	33437	27130	00065	59418	15906	91072	67333
26568	19254	82207	36500	38441	56842	32589	84476	79303	86435
04011	49424	75538	64977	99870	26805	65272	19759	47046	56612
93594	20929	17240	21175	81747	90382	63581	03616	98508	25243
69545	88610	58152	32517	36099	51764	54186	17151	59213	95478
64721	98756	66448	67183	23934	65269	17287	59922	13980	71915
36057	93350	95692	28285	10104	08439	37326	06291	11418	49153
90996	23811	26720	88827	10274	41861	49863	30388	38655	65462
65609	69896	05123	74910	36245	88438	14090	75597	24444	71431
23358	03025	69232	36779	88012	70912	96466	35093	55460	96367
52614	95001	90328	21395	07002	21949	17036	39063	50830	51137
99774	41743	96146	68469	96182	12571	31298	93765	78771	21119
90760	81062	71851	23532	63135	31606	17033	00200	39907	50954
24141	46268	20135	84542	34289	40176	69003	34722	90922	60473
63764	91862	03570	62677	29711	21124	35238	24312	00505	61459
42746	94973	60950	98208	05294	29281	19447	98082	02629	34875
94943	15436	72598	92030	24148	98475	72310	10217	22851	03464
35178	23245	05852	57799	29886	68913	73634	29197	25923	57797
58413	11237	37680	15634	34987	46148	10421	65615	47622	26969
14456	16883	89827	22769	48466	33088	57050	93757	41804	91991
31118	61985	23392	10138	93026	38853	74967	62457	01183	64565
61746	90088	67561	81974	86527	30420	33162	07309	65596	34823
57790	13297	37131	23058	16725	76932	32166	04793	11131	39160
74067	04314	00701	30028	75095	54702	43649	92736	48763	74530
46405	33295	33755	82716	68837	86484	58271	10998	48022	87465
66472	82819	47306	66733	63900	22551	97607	82654	09841	65968
53835	12242	35989	95876	58703	07270	60377	92824	55411	94934
74205	92013	03159	38446	24673	04645	57147	16992	94985	97902
48575	05782	84329	75012	64616	56010	87918	55166	28559	74878
53552	76945	79498	78613	05586	61380	07334	52687	56121	25848
35321	88669	30156	66583	59315	20750	73504	20750	77690	95685
50818	51092	95685	11839	48726	28553	34120	60377	92824	55411

Answers to Exercises

Chapter 2

2 Mean for raw data is 129.7 persons/km^2. No, the mean is too susceptible to extreme values, so the median = 33.5 is preferable. No, the latter quantity is the total population divided by the total area.

3 *A*, mean 3.37, median 3, mode 2; *B*, mean 3.07, median 2, mode 2; *C*, mean 2.69, median 2.5, bimodal 1 and 2.

4 5 and 21 clearly incorrect, 12 is actual value; 11 and 17 clearly incorrect, 10 is actual value; 1,045 clearly incorrect, actual value is 1,149.

Chapter 3

1 (a) Standard deviation is 469.7 (with a mean of 129.7).
(b) Mean 69.6, standard deviation 83.2. Illustrates how both measures are susceptible to extreme values, so in this case they are of rather limited descriptive usefulness.

2 *A*, 1.52; *B*, 1.65; *C*, 1.39.

3 *A*, 69–100; *B*, 60–68; *C*, 56–9; *D*, 50–55; *E*, 42–9; *F*, 31–41; *G*, 0–30.

4 Mean, 8.89 months; standard deviation, 3.00 months.

Chapter 4

1 (a) $Pr(E_1 E_2) = 0.7 \times 0.8 = 0.56$; (b) $Pr(\bar{E}_1 \bar{E}_2) = 0.3 \times 0.2 = 0.06$;
(c) $Pr(E_1 \bar{E}_2) = 0.7 \times 0.2 = 0.14$; (d) $Pr(\bar{E}_1 E_2) = 0.3 \times 0.8 = 0.24$;
(e) $1 - Pr(\bar{E}_1 \bar{E}_2) = 1 - 0.06 = 0.94$.

2 (a) $Pr(\text{no boy}) = (\frac{1}{2})^4$; (b) $Pr(\text{2 boys and 2 girls}) = \binom{4}{2}(\frac{1}{2})^4 = \frac{6}{16}$;
(c) $Pr(\text{at least one boy}) = 1 - Pr(\text{no boy}) = 1 - (\frac{1}{2})^4$;
(d) $Pr(\text{no boy}) + Pr(\text{1 boy}) + Pr(\text{2 boys}) = \frac{11}{16}$.

3 *Binomial distribution* (a) $Pr(\text{at least 2}) = 1 - Pr(0) - Pr(1) = 1 - (0.83)^{10} - \binom{10}{1}(0.83)^9(0.17) = 0.527$;
(b) $\binom{10}{3}(0.28)^3(0.72)^7 = 0.264$;
(c) $(0.45)^{10} + \binom{10}{1}(0.55)(0.45)^9 + \ldots \binom{10}{5}(0.55)^5(0.45)^5 = 0.496$;
(d) $Pr(\text{1 or 0 single person}) = (0.72)^{10} + \binom{10}{1}(0.28)(0.72)^9 = 0.183$.

4 *Poisson distribution* (a) $Pr(0) = e^{-4.7} = 0.009$; (b) $1 - Pr(0) - Pr(1) = 0.948$;
(c) $Pr(0) + \ldots Pr(6) = 0.805$; 4 with a probability of 0.185.

5 *Binomial* (a) 0.040, (b) 0.091, (c) 0.180; *Poisson* (a) 0.042, (b) 0.092, (c) 0.184; *normal* (a) 0.052, (b) 0.080, (c) 0.184. The Poisson gives a reasonably close approximation to the binomial values for (a), (b), and (c), while the normal is only close in the case of (c). The normal approximation is not appropriate in this case.

Chapter 5

1 $\bar{x} = 12.798$, $s = 7.204$ (with the open-ended interval taken as having length 4); 95 per cent limits are $12.798 \pm 1.96 \times 7.204/\sqrt{400}$, i.e. 12.09 and 13.50. For this large n, it is appropriate to use z- rather than t-values.

2 $\bar{x} = 8.275$, $s = 5.729$; 99 per cent of area under t-distribution with $\nu = 39$ is between $t = \pm 2.708$; 99 per cent limits are $8.275 \pm 2.708 \times 5.729/\sqrt{40}$, i.e. 5.82 and 10.73.

3 $r/n = 0.34$; 95 per cent limits are $0.34 \pm 1.96\sqrt{(0.34 \times 0.66)/250}$, i.e. 0.281 and 0.399.

4 Equate $1.96\sqrt{(0.2 \times 0.8)/n} = 0.02$ from which $n = 1536.6$, so a sample of at least 1,537 is required.

Chapter 6

1 $Pr(4) + Pr(5) + Pr(6) = 0.0013$; reject H_0 at 1 per cent level (one-tailed test).

2 $Pr(2) + Pr(1) + Pr(0) = 0.143$; do not reject H_0 at 5 per cent level (one-tailed test).

3 $z = 2.54$; reject H_0 at 1 per cent level (one-tailed test).

4 $z = 2.72$; no.

5 $t = -1.57$, $\nu = 15$. Do not reject H_0 at 5 per cent level in two-tailed test, since critical t-values are ± 2.131.

Chapter 7

1 (a) $z = -3.20$; reject H_0 at 1 per cent level (two-tailed test).

(b) $t = -2.99$, $\nu = 28$; reject H_0 at 1 per cent level (two-tailed test). $F = 1.562(5)$; $\nu_1 = 15$, $\nu_2 = 13$. F-ratio is less than critical value at 5 per cent point, so do not reject hypothesis of homogeneity of variance.

2 $t = 2.29$, $\nu = 8$; reject H_0 at 5 per cent level (one-tailed test). Test requires interval scaling with population differences normally distributed.

3 $z = 1.89$; do not reject H_0 (two-tailed test).

4 $U = 9$ which is below critical value of 13, so difference is significant.

5 (a) $D = 0.27$; this is less than $225/(24 \times 25) = 0.375$, so do not reject H_0 at 5 per cent level.

(b) $D = 0.27$; this is greater than 0.147, so reject H_0 at 1 per cent level.

Chapter 8

1(a) $F = 10.69/0.942 = 11.35$; with $\nu_1 = 3$, $\nu_2 = 21$ critical value at 1 per cent level is 4.87, so result is highly significant.

(b) $H = 17.16$; critical value of χ^2 with $\nu = 3$ is 11.35 at 1 per cent level, so result is again highly significant.

2 No; $s_b^2/s_w^2 = 50.8/54.7 = 0.93$.

Chapter 9

1 $\chi^2 = 14.31$; with $\nu = 6$ critical value is 12.59, so reject H_0.

2 Exact probability of obtained outcome or one more extreme is 0.032; reject H_0 at 5 per cent level.

3 $\chi^2 = 2.376$ (corrected for continuity); with $\nu = 1$ critical value is 3.841 at 5 per cent level, so do not reject H_0. $\phi^2 = 0.045$.

4 $\chi^2 = 257.4$; with $\nu = 9$ critical value is 21.67 at 1 per cent level, so reject H_0. $C = 0.493$; $T = 0.327$ ($T^2 = 0.107$).

5 $\chi^2 = 9.50$; with $\nu = 7$ critical value is 18.48 at 1 per cent level, so do not reject H_0.

Chapter 10

1 $y = 76.335 - 2.393x$; $r = -0.840(5)$. When testing r for significance, calculated t-value is -6.58, which is below critical value of -2.101 in two-tailed test with $\nu = 18$, so null hypothesis $\rho = 0$ may be rejected. There is evidence that poorer examination marks tend to follow greater time spent on non-academic activities.

2 $y = 13.217 + 0.378x$; $r = 0.870$. The high r-value has the appearance of a 'spurious' correlation generated by a trend away from antenatal examinations in clinics during a period when general antenatal health care was improving, leading to a lower perinatal mortality rate.

3 $r = 0.133$, a low value suggesting that during the period as a whole house-building was not dominated by factors which were equal in their impact on the two sectors. No doubt new governmental initiatives prompted the high level of public-sector activity after 1964 and its decline after 1970.

4 $r_s = -0.8773$; calculated t-value is -6.59, which is below critical value of -3.852 in a one-tailed test with $\nu = 13$, so result is highly significant.

Index

accuracy 36, 72
advance estimates 34–6, 42, 86–7, 152
age 6, 13–15, 23
Analysis of variance 121–7
Annual Abstract of Statistics 169, 172
approximation 36
arithmetic mean, *see* mean
attributes 6
averages 1, 16–23

bar charts 9–10
bias 72, 75, 99
Binomial distribution 59–65
  general term 60
  mean 62
  and normal curve 62–5
  and Poisson 65–7
  and proportions 82–3
  variance 62
Binomial test 91–3, 98

calculators 22, 31, 36, 137
Census of population 1, 170–1, 176, 183
  of 1971 8, 167, 171
  weakness of 171
central limit theorem 73
Central Statistical Office 183
chi square test 129, 131–7
  correction for continuity 135
  degrees of freedom 133, 144
  for 'goodness of fit' 141–4
  restrictions on use 134–5, 136–7
  statistic 133–4, 139
  table for 188
combinations 56–8, 59
confidence intervals 76
  and hypothesis testing 101–2
  for mean 76–82
  for proportion 82–4
confidence limits 76, 77–8, 80–1, 83–4
contingency coefficient *C* 140–1
contingency problems:
  measures of relationship 138–41
  significance tests 131–8
  tables 131–2, 135–8
continuous variables 6, 11–16, 18, 23, 36
correlation 2, 154–61
correlation coefficient:
  Pearson product-moment 154–6
  Spearman rank 159–61
covariance 155
criminal statistics (sources) 175
critical region, *see* rejection region
critical values 95, 96–7

data, types of 4–7
deciles 27
decision making 89–91, 92, 95
degrees of freedom 79
  for chi-square 133, 144
  for *F*-distribution 108
  for *t*-distribution 80
  of variance estimate 122–4
descriptive statistics 1–2
Digest of Welsh Statistics 161, 169
discrete data 10–11, 19
discrete variables 6, 15, 18, 36
dispersion, *see* variation
distribution free tests 112

education statistics 168, 176–83
EEC statistics 170
electoral roll 4, 70
errors 36, 90–1
estimation 3–4
  of mean 75–82, 86–7
  point 76–7, 83
  of proportion 82–6
  in regression 151–2, 153
  of variance 79
  see also advance estimates
estimators 72
  unbiased 79
events 45–6, 48
  compound 51–3
  independent 50, 51, 52
  mutually exclusive 54
  rare 67
expected value 62, 72, 79, 83
exponential model 156–8

Facts in Focus 169–70
F-test 108, 124
  tables for 189–90
factorials 55–6
Family Expenditure Survey 172
finite population correction 73–5, 83, 98, 99
Fisher exact test 137–8
frequencies 9

relative 10, 13
frequency distributions 8–16, 20–23
cumulative relative 115–17
relative 10, 13–14
frequency tables 9–11
grouped 11–16

gaussian, *see* normal distributions
General Household Survey 172, 176
'Goodness of fit' test 141–4
graphical methods 9–15, 148–50, 153–4
grouping of data 11–13, 22–3, 142
size of intervals 15
Guide to Official Statistics 167–8

Health Statistics (sources) 175
histograms 11–13
housing statistics (sources) 175–6
hypotheses:
types of 90
hypothesis testing 4, 89–91, 92–3
normal curve methods 94–7
and confidence intervals 101–2

income 6, 15–16, 19
independence:
of events 50, 51, 52
of random samples 105
statistical 141
inferential statistics 2–4
intelligence quotients 39–42, 81
intercept 150
interval scales 5–6, 7

Kolmogorov-Smirnov test 115–17
table for 192
Kruskal-Wallis test 127–9
table for 193

least-squares criterion 150–2
level of significance 90–1
linear models 150
line charts 9
line of best fit 150–4
lottery method 70

Mann-Whitney *U*-test 112–15
table for 191
matched pair design 109–10, 117
Maunder, W. F. 168
mean 18–23
calculation of 20–23
confidence intervals 76–82
and mode 42
as point of balance 18–19
of probability distribution 60–1
sampling distribution 73–5
tests for differences 104–10, 125–7
tests on 98–101
measurement 5–7
measures of central tendency 16–23
measures of relationship 138–41
median 16–18, 27
mode 16
and mean 42
models 149, 162–3
linear 150
exponential 156–8
Monthly Digest of Statistics 169

nominal scales 5, 7, 16
tests involving 91–4, 97–8, 110–12, 131–8
nonparametric tests 112
normal curve 37, 38
fitting 141–4
standard 39–40
normal distributions 37–42
and hypothesis testing 94–7
standard form of 39–42
table for 185
normal equations for least squares 152

one-tailed tests 96–7
opinion polls 2
ordinal scales 5, 7, 16, 174
tests involving 112–17, 127–9, 160–1

parameter 3
percentiles 27
perfect association 139–40
Phi-square coefficient 139–40
point estimate 76, 77
Poisson distribution 66, 67
approximating binomial 65–7, 93–4, 98
population 2
population statistics 157–8, 166, 167, 170
precision 72, 75–6, 78
and sample size 86–7
probability:
addition rule 54
and combinations 54–8
conditional 49–50
definition 45–8
product rule 54
and relative frequency 46–8
probability distributions 58
binomial 59–65

mean of 60–1
normal 65
Poisson 66–7
variance of 60–1
proportions
confidence intervals 82–4
estimating 82–7
tests for differences 110–12
tests on 97–8

Quartile deviation 27
quartiles 27
quasi-random sampling 71–2

random numbers 70
table of 183–4
random sampling 69–71
simple 69–70
stratified 71
range 26–7
and standard deviation 42
rank correlation 158–61
rating lists 70
ratio scales 5–6
regression analysis:
purposes 147–9
rejection region 90, 94–7
repeated trials 46–8
research design 4, 85–7, 109, 117, 121, 166, 183

sample 2
size 78, 85–7
sampling:
distributions 72–5, 83
error 4
fraction 71
frame 70, 71
methods 49, 52, 69–72
scales 4–7
scatter diagrams 148–9, 152, 153–4, 162
school leaving qualifications 177–8
sigma notation 19–20
significance levels 90–1
Skewness 42–3
small sample tests 106–7, 137–8, 156
social class categorisation 173–4
Social Trends 168–9, 176–9, 180–2
socio-economic groups 173–4
standard deviation 30
adjusted 79
calculation of 30–6
and normal distribution 42
and range 42
standard error:
of difference of means 105–6
of difference of proportions 111
of mean 73, 99
of proportion 83
Standard Industrial Classification 173
Statesman's Year-Book 170
statistic 1, 3, 72
statistical inference 2–4
statistical methods 1–4
Statistical News 168
statistics:
descriptive 1–2
inferential 2–4
meaning of 1
stratified random samples 71

T-distribution 80, 100
and normal distribution 80–1, 100
table of, 187
use of 80–82, 100–1, 106–8
trends 157, 172–3
Tschuprow's *T* 140
two-tailed tests 95–7
type I and type II errors 90–91

United Nations publications 170

variables 6
variance 29–30
estimates 79, 121–7
variation 19, 26
measures of 26–36

Z-values 39–42
table of areas 185–6